D0524500

Rational Methods, Prudent Choices:
PLANNING US FORCES

Rational Methods, Prudent Choices:
PLANNING US FORCES

Robert P. Haffa, Jr.

1988

NATIONAL DEFENSE UNIVERSITY PRESS
Fort Lesley J. McNair
Washington, DC 20319-6000

National Defense University Press Publications

To increase general knowledge and inform discussion, NDU Press publishes books on subjects relating to US national security.

Each year, in this effort, the National Defense University, through the Institute for National Strategic Studies, hosts about two dozen Senior Fellows who engage in original research on national security issues. NDU Press publishes the best of this research.

In addition, the Press publishes other especially timely or distinguished writing on national security, as well as new editions of out-of-print defense classics, and books based on University-sponsored conferences concerning national security affairs.

————————————

The manuscript of this book was proofread and indexed by SSR, Inc., Washington, DC under contract DAHC 32-87-A-0013.

NDU Press publications are sold by the US Government Printing Office. For ordering information, call (202) 783-3238 or write to the Superintendent of Documents, US Government Printing Office, Washington, DC 20402

Library of Congress Cataloging in Publication Data

Haffa, Robert P.
  Rational methods, prudent choices.

  Bibliography: p.
  Includes index.
  1. Military States—Armed Forces—Organization. 2. Military planning—United States. I. Title. UA23.H25   1988   355.3′0973   88-29019

First printing, December 1988

For
Colonel Robert P. Haffa, Sr., USA (Retired)

CONTENTS

ILLUSTRATIONS

FOREWORD

Planning for US military forces goes on regardless of the political party in power, the state of the budget, or the issues of the moment. Because planners decide the size and shape of land, air, and sea forces, force planning is at the very core of our national security effort.

In this primer on force planning, Colonel Robert P. Haffa reviews the process used to structure our strategic, general purpose, and rapidly deployable forces. He contends that many people both within and outside the defense community do not fully understand force planning methods. Too often, he writes, military planners themselves—caught up in the daily pressures of the bureaucracy—focus on parochial, near-term issues. At the national level, far too many public debates are cast in terms of dollars instead of national objectives, missions, and forces. Haffa calls for a return to first principles, recommending these four guidelines for force planning: emphasize coherent policy relationships; rely on empirical data; stress planning, not budgeting; consider the long term.

Haffa shows that rational planning methods lead to prudent choices. His analysis reminds force planners never to lose sight of fundamentals, especially while prodding national leaders to pay attention to the rational methods of force planning. This fresh study of how we plan our military forces inspires us to get back to the basics essential for informed, productive debate on defense issues.

BRADLEY C. HOSMER
Lieutenant General, US Air Force
President, National Defense
University

PREFACE

As an officer in the United States Air Force, I have been fortunate enough over the last ten years to study the planning of US military forces. My education in force planning issues began when I was a graduate student at the Massachusetts Institute of Technology where I was sponsored by the Air Force Academy's Political Science Department, under the auspices of the Air Force Institute of Technology.

When I returned to the Academy to teach political science, I noted the absence of a force planning approach in our course on American defense policy. Although the department there had pioneered the teaching of defense policy at the undergraduate level with its text *American Defense Policy*, that volume, through a number of editions, had become largely strategic in focus and bureaucratic in flavor. We seemed to be teaching our lieutenants-to-be that American defense policy was an arcane subject conceptualized by intellectuals, manipulated by politicans, and driven by budgets. Under this paradigm, the planning of US military forces—determining the number and type of army divisions, air force wings and naval battle groups—resembles the irrational outcome of a bureaucratic political process. Neither explanations of the rational basis for these existing forces nor presentations of the quantitative methodology available to and used by force planners are given any emphasis.

This book aims to change that approach. Students of defense policy need to know that current force planning has indeed been based on rational methods and prudent choices. I hope the editors of a new version of *American Defense Policy* reflect that reality, and are encouraged through this work to include force planning within the Academy's defense policy curriculum.

This book, then, is not intended primarily for those positioned to influence defense and force planning decisions in the near term. Nor is this primer on force planning addressed to the

Pentagon analysts who use tools and techniques far more sophisticated than those sketched here. I send them applause, not advice. But many major defense issues are never thrust under the scrutiny of those expert analysts. All too often force planners on various service staffs lose their view of the forest by concentrating on a few leaves. Thus rational methodology can indeed be entangled in the thickets of bureaucracy and the thorns of the budget.

I am convinced that an understanding of force planning has been lacking by many who were debating defense issues in the 1980s. Why 600 ships? Why 100 B1-Bs? 50 (or 100) Peace-keepers? Why 6 divisions in Europe? How do we continue to accomplish an essential set of missions with less and less real money? No defense decisions are more important than these. They demand that planners not surrender—especially not sur-render *en masse*—to the easy temptation and momentary satis-factions of parochial views. In talking with many of the participants in the planning process, I have grown convinced that they follow the wrong approach to force-structure decisions only because they are unaware or distrustful of the rational methods that make prudent choices consistently possible. To avoid these no longer useful and progressively more dangerous approaches—marginal-adjustment, cost restructuring, and pro-gram stretch-outs—they need to understand more clearly the matters treated by this book. If that exposes this book to labels like "didactic," so be it. I want to return to first principles, to push back from the conference table and the in-box and reflect, to refresh our understanding of the history of force planning so that we make better decisions in the thundering present.

I wrote this book during my year as a student at the National War College and as a Senior Fellow at the National Defense University. Throughout, I received help from a number of people to whom I owe my thanks. Much of what is written here relies on the previous work of my mentor at MIT, Pro-fessor William W. Kaufmann. His former students will quickly recognize that intellectual debt. Included among those who took time to read and comment on the draft manuscript were Lynn Davis, Dick Fast, Bob Kennedy, Roy Stafford, and Perry Smith. Patient and persistent editors at NDU Press worked to bring this book to print. Many others—in Cambridge, Colorado Springs, and Washington—contributed, but the final responsibility is, of course, mine.

ROBERT P. HAFFA, JR.
Fairfax, Virginia

Rational Methods, Prudent Choices:
PLANNING US FORCES

1. A PRIMER ON FORCE PLANNING

Since the end of 1979, with its dual crises of US hostages in Iran and Soviet troops in Afghanistan, American defense policy has once again become the object of public concern, academic interest, and government effort. The reluctance to use the military as an instrument of US foreign policy lasted a decade. Effects of the American withdrawal from an unsuccessful military engagement in Vietnam included a retrenchment in Southeast Asia, a reduction in the defense budget, and a rollback in US military capabilities.

But great powers cannot so easily forswear their political responsibilities and military commitments. Thus, when US global interests became threatened in the late 1970s, the return to a military option came sooner than many expected. The president who entered office dedicated to reducing force deployments and defense spending left office emphasizing anew the military instrument and declaring an American willingness to use force in regions far from the US mainland.

In 1981, that proposed defense buildup gained both momentum and money. Though a stagnant industry in the 1970s, defense became the growth industry of the 1980s. (See Figure 1:1.) Unsurprisingly, the direction federal funds flow tends to attract attention in America. Some line up for their fair share; others question the course being set. The debate surrounding American defense policy in the 1980s featured several contradictory pairings: quality versus quantity, attrition versus maneuver, strategists versus managers. The debates have been complex, the results inconclusive.

Throughout the early 1980s, the military buildup continued. Events in Southwest Asia contributed to a permissiveness in public opinion allowing increased defense spending for

Figure 1:1
Defense Budget Authority and Outlays, Fiscal Years 1950–1992
(in billions of 1988 dollars)

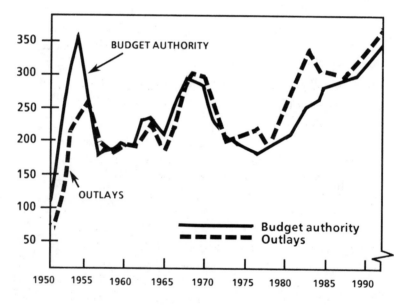

SOURCE: Unpublished material furnished by the Office of Management and Budget. The numbers for 1987–1992 reflect the President's fiscal 1988 budget request.

greater military capability. However, by the mid-1980s, growing budget deficits gave pause to the planning of forces and the procurement of weapon systems. These criticisms of US defense policy based on the budget deficit are not only often misplaced, they also fail to consider the fundamentals that underlie the planning of our military forces. This primer seeks to foster understanding of force planning basics by focusing on the following:

The baseline force: What are current force levels? How were they reached?

The adequacy of the force: How capable are current US forces of meeting anticipated contingencies? How can we test those capabilities?

The future of the force: If deficiencies are demonstrated, on what basis should the United States plan its forces to remedy those deficiencies?

An answer to the first question calls for a description and explanation of the basis for the planning behind existing US military forces. In a time when force planning is too often thought of in relation to some percent of the gross national product, this methodology is frequently overlooked. Explaining this methodology highlights the rational methods and prudent choices used to construct the baseline force. The central thesis of this study answers the second question. A rational framework for planning military forces based on tests of their adequacy—threat assessment, campaign analysis and quantitative modeling—exists and has been used effectively. A realistic conclusion addresses the third question: methods used successfully in the past should not be carelessly cast aside as the United States embarks on a major military improvement program—but reemphasized as balance is restored to the nation's budget.

Although the fundamentals of force planning are admittedly incomplete, they are all too often forgotten in analyses that stress bureaucratic outcomes of defense decisionmaking and international perceptions of force capability. Today, far too many defense debates are cast in terms of dollars rather than in terms of objectives, missions, and forces. My purpose here is to argue that a more satisfactory method of understanding the base-line force and evaluating programs to improve it is to estimate the extent to which existing and planned forces can meet national objectives and commitments. Planned increases or reductions in those forces must be related to their capability to meet those goals.

What Force Planning Is

Force planning is subsumed under defense policy, which in turn acts in support of United States national security policy and foreign policy. One of the best ways to distinguish between force planning and other elements of defense policy is to differentiate among policy levels. This approach is not new. Writing in *Foreign Affairs* in 1956, Paul Nitze distinguished between declaratory policy—statements of political objectives with intended psychological effects—and action or employment policy—concrete military objectives and plans employing current forces in support of those objectives.[1] Nitze also saw the requirement to match the two levels closely, lest declaratory policy appear hollow or employment capability inadequate. But that fit has never been perfect.

Nitze's concept has since been refined. Donald Snow and others have inserted a policy level between declaration and employment: force development and deployment.[2] Force planning is the development of forces flowing from the requirements of declaratory policy or the shortfalls in employment policy. Force development planning should, therefore, unite a declared strategy and the means to implement it. Snow also noted some important differences between the levels. Declaratory policy is the strategy of the elected political leaders. Therefore, it is the most political level, subjected to cosmetic if not comprehensive change during election cycles. Declaratory policy is also the level at which most academic debate is focused; it is the most public decision level. Employment or action policy, on the other hand, lies in the domain of the military. It demonstrates remarkable continuity, even in the face of political change. Often it is highly compartmentalized and protected from public view; strategic nuclear employment policy is the most secret decision level.

What of force planning? A split exists in this middle king-dom of decision levels. Those with a micro-perspective on force planning tend to concentrate on weapons system acquisition. Case studies abound in documentation of the difficulties of weapons system development. The majority of these analyses ultimately explain the acquisition process as a nonrational, polit-ical-military-budgetary compromise. This, of late, has been the most maligned decision level.

But the macro-perspective on planning US military forces may be the most ignored decision level. The concern at this level is not with what individual weapons systems are procured but what military forces are required to meet specific contingen-cies. The units of analysis are not single M-1 tanks or F-111 air-craft but collective forces of strategic missiles, army divisions, or carrier battle groups. Judgments are required not only on the size and structure of the force, but also on the mix of force modernization, readiness, and sustainability. Force planning must be related to declaratory and employment policy in a rational way. This study assesses the importance of such an approach to force planning.

Given these three policy levels—declaratory, force development, employment—what would a rational force plan-ning process look like? While there is an obvious difficulty in attempting to link these levels, an ideal process can be designed. The declaratory policy should come first: incorporat-ing objectives formulated by political leaders enjoying popular support. The employment policy should follow: utilizing exist-ing forces to accomplish the declared strategy. Force planning is third: developing forces in support of declared policy and designing forces to overcome shortfalls in contingency war plans. Finally a budget emerges: within given constraints, sup-porting the planned and programmed force.

Even dilettantes of defense policy who do little more than glance at newspaper headlines will recognize that the current force planning process does not work this way. There is not a close link between declaratory and employment policy; there is, instead, a great gap. To a certain extent, this gap will always exist. The danger is that, in a strategy-force mismatch, the choices too often become stark: reduce US commitments abroad or expand defense budgets to meet all comers. When defense policy is reduced to such simplistic terms, force planning becomes programming. Programming ends with a budgetary battle in which arbitrary across-the-board cuts are made to fit fiscal guidance fashioned by committee. This emphasis on defense spending cannot be characterized as a rational way to plan US military forces.

There is another way. It has been applied and applied successfully. The baseline force in both strategic nuclear and general purpose forces derives from rational methods of contingency planning. When tested with these methods and models, much of the baseline force appears adequate. Planners may reach prudent choices by employing these rational methods when they consider improvements to documented deficiencies in the force structure. Programmers may also wish to return to these methods as they seek more modest defense growth in the late 1980s and 1990s.

What Force Planning Is Not

If the rational chain of the force planning process is continued, the result should be a budget. Since 1961, the Defense Department has constructed its budget through the Planning, Programming, and Budgeting System, referred to as the PPBS. Force planning is not the PPBS, and this study does not attempt to explain that system.[3] Indeed, Washington wags have suggested that the first "P" in PPBS is silent. They may be close to the truth. In theory, knowledge of the PPBS helps to understand the force planning process and its need for rational input.

The planning cycle is actually the longest of the series of events in the PPBS. Joint and service staff force planning starts as much as a full year before the initial development of the OSD-directed Defense Guidance. Thus there is plenty of time for planning and analysis. But a major disconnect occurs within this process. The service and joint force planners submit a "planning force"—levels required to meet declared policy with reasonable assurance of success. The Defense Guidance responds with fiscal constraints to be used by the services in preparing their programs.

Consequently, the Defense Guidance ends the planning process and turns the attention of would-be defense planners to programs and budgets. The concept of moving from a minimum risk force to a constrained force can be valuable if accompanied by a rational process to consider and make explicit the political cost of increasing risk as the dollar cost declines. But there is no meaningful coordinated joint force planning process that makes trade-offs within these areas. Instead, we usually find a bureaucratic battle of service-oriented programs and across-the-board budgetary cuts, revising, as Lawrence Korb writes, the would-be rational policy of moving from declared goals to forces to funding.[4]

The actual process is too familiar. The administration decides on the size of the total federal budget and allocates some of that to defense. Further distributions apportion money required to execute the program among the services (called Total Obligational Authority, or TOA). Programming takes place based on allocation. Planning is left behind. Such an emphasis on funding focuses on "how much" and disregards the important questions of "what for, how many, and how well." Thus, once the planning cycle has been put to rest, the real business of defense takes over. The services dominate the programming cycle with their separate roles, missions, and agendas. Here decisions are made, not only on weapons programs, but on what kind and on how many forces can be acquired based on service desires and resource limits.

In the budgetary phase, the President and the Office of Management and Budget review the programs passed on by the Secretary of Defense. Ultimately, the Congress gets several shots in what has by now become merely a fiscal exercise. First a budget resolution establishes a ceiling on expenditures. Congress then embarks on a lengthy and multi-committed exercise to authorize and appropriate the funds. It micro-manages the remnants of the planning process.

By the time Congress has matched its target with its final budget, the planning force, or some rational version of it, has been forgotten. Although the final budget may vary considerably (higher or lower for separate line items, depending on the political and economic climate) from the President's budget, the alterations often demonstrate little consideration of those forces planned against assessments of enemy threats. Rather, these changes in expenditures are often based on constituent interests and short-term political perceptions, and often evidence a tendency to reduce the "fast money" of pay, allowances, and operations and maintenance costs rather than to cut "big ticket" items. Thus programming, primarily, and budgeting, ultimately, rule the PPBS. Rational force planning, a realistic matching of national commitments and forces based on threat assessment and resource constraints, is unachievable in a system dominated by parochial interests. Fiscally constrained strategy does not have to be developed; it is already a reality.

Parenthetically, force planning is not simply systems analysis, although analysis is an important part of rational force planning. Systems analysis-bashing is considered good sport these days, principally by agencies and individuals feeling threatened by force and budgetary analyses emanating from the Office of the Secretary of Defense.[5] These laments frequently culminate in the accusation that systems analysts are closet strategists who, given any encouragement at all, will eagerly fill the void between declaratory and employment policy. Analysts have, at times, played that role, but probably more because other actors abdicate that function than because of their own ambitions. Most analysts would agree that they should not make decisions and should not set goals or assign priorities. But analysis can assist the force planning decision process.

Analysts pursuing rational force planning as a link between declared policy objectives and their implementation will also admit to shortcomings. At times, economic graphs have been allowed to sweep over less quantitative political judgments and military expertise. Analysis has been used in pursuit of political goals as well. But analysts also will argue that their influence is less than they desire or their opponents fear.[6]

From this parenthetical discussion, a larger question emerges: what could substitute for analysis in the systematic planning of military forces? Some argue for military experience, yet force planners assert that decisions must be a judgmental blend of analysis and experience with civilian control imperative. Others argue that analysis omits important, unquantifiable variables. Yet there are indices, imperfect but important, that can weigh the training, morale, and maneuver of opposing forces over varying terrain. The most difficult issue to reconcile is that of perception. Notions of strategic superiority or conventional parity have great political appeal but little meaning to the force planner. Static measures of comparison and fuzzy ideas of how forces are perceived by allies and adversaries are poor guides to rational force planning, and likely guarantors of unlimited budgetary claims.

Planning US Military Forces

Rational force planning is an analytical process designed to link declaratory and employment policy. To that end it assesses the military balance in possible contingencies, measures force capabilities in relation to requirements, and, after cross-program evaluation, establishes broad priorities for allocating resources. The task is big. But planning is essential in determining whether the US defense effort coherently supports US goals. Without planning, the larger process of formulating defense policy fragments into separate, uncoordinated, and irrational programs.

This book is meant to be a primer on force planning. The process is complex. Describing and explaining that process, and attendant arguments for rational relationships among decision levels, is made more difficult by the number of foreign policy commitments the United States has undertaken and by the variety of military forces that the defense of so many agreements requires. We must, therefore, examine the planning of US military forces under broad headings.[7] I have selected three.

Chapter 2 looks at the first, planning the strategic nuclear force. Historically, these forces have been planned separately from the general purpose forces owing to the uniqueness of nuclear weapons, the uncertain situations in which they would be employed, and the bipolar nature of a strategic confrontation. The explanation of the planning of US strategic forces will follow the model suggested earlier—the decision levels of declaration, planning, and employment. But the strategic nuclear force has encountered great difficulties uniting declaratory and employment policies. I will focus briefly on the planning of the baseline strategic force: the "triad" designed in the 1960s. What determined the numbers of 1,000 Minutemen and 41 Polaris submarines? How rational was that force planning process? What can we learn from that planning effort to aid us in planning for the 1990s and beyond?

Chapter 3 asks and answers similar questions based on the same framework but focused on the second heading—general purpose forces. In this case the declared policy consists of US treaty commitments abroad. Here the strategy-force mismatch is more apparent. But the method of contingency analysis, albeit more complicated than in the strategic nuclear equation, still has value. The United States planned its general purpose forces based on a strategic concept of the kinds and numbers of wars it might be called on to fight—the so-called "2½ war" and "1½ war" models. Analysis helped determine the number and size of army divisions, tactical fighter wings, and combat ships required to fight these wars in specific contingencies. These methods appear to retain relevance for contemporary force planning.

Chapter 4 examines force planning for rapid deployment. Here rational planning has been less successful because of a failure to procure sufficient strategic mobility forces to support limited contingency operations rather than of problems in the planning process. The strategic concept was also faulty, however, and early organizations designed to support a limited contingency, envisioned as directing multi-contingency forces, proved incapable of operating in the real world. The planning of a coherent limited contingency force, whether from a revised strategic concept or merely from the larger defense budgets of the early 1980s, appears to be progressing. Will this progress continue with the leaner budgets to come?

Chapters 2 through 4 examine the ghosts of force planning past, concentrating on the period 1960-1980. Chapter 5 examines present and future force planning, questioning how lessons learned are being applied. The early chapters suggest that a rational basis for the design of US military forces does exist, has been used, and has produced a baseline force tested against plausible contingencies. The final chapter inquires into the success of current force planning and suggests that newly imposed constraints on defense budgets may call for a return to the rational methods and prudent choices of the past.

Although this book seeks to make a case for the application of analysis to force planning in a rational process, it also seeks a middle ground between contending theses and antitheses. One of the dialectics attending the defense debate is the dichotomy between experience and rationality or, as defense analyst Edward Luttwak has coined it, between "military effectiveness" and "civil efficiency." The argument against rational analytic methods asserts that, paradoxically, they do not reflect reality since there are simply too many variables and too many unknowns in the fog and friction of war. The middle ground is that both experience and reason are required. Churchill's dictum that "it is sometimes necessary to take the enemy into consideration" applies. But enemy considerations are difficult to assess. The more problematic decision is how much attention to give both analysis and experience.

Other sets of opposing forces act as *leitmotifs* in this force planning debate. In *The Politics of the Budgetary Process*, Aaron Wildavsky called the budget the "life blood of the government." In *The Air Force Plans for Peace*, Perry Smith termed doctrine "the life blood of the service." Much of both types of blood gets spilled in annual bureaucratic combat; there has even been some practice bleeding. The flow can be stanched and transfusions obviated if service budget allocations become rational outputs of the system rather than priority-setting inputs.

A third issue is one of management style within the Defense Department, often framed in arguments for centralization or decentralization. There have been definite swings in management style—more centralized under Secretaries McNamara and Brown, more decentralized under Secretaries Laird and Weinberger. Many observers contend that the Defense Secretary's reliance on his staff of systems analysts to aid defense decision making has removed force planning power from the services, the commanders-in-chief (CINCs) and the Joint Chiefs of Staff. But the services have their analysts, too, and there are advantages and disadvantages to each style. Some students of the issue have suggested that the management of defense is likely to depend on the political and military climate; others have posited that the structure of decision making makes little difference.[8] Such contentions aside, only a strong Secretary of Defense can make the force planning process work well. He ultimately makes the hard choices on force development, regardless of the managerial style used to generate options.

A fourth theme that runs through this force planning study is the conflict between the realities of the short term versus the implications of the long term. Lord Keynes warned us that the long term ceases to compel attention when the out-years begin to exceed our life expectancy. A political corollary to Keynes' dictum contends that the long term fails to compel when it exceeds the watch of those in power. But force planners, owing to the long lead times of weapons acquisition, must reject such corollaries. The tyranny of annual cycles in formulating budgets, programs, and even strategies pressures planners to perform in the present. Force planners are also susceptible to an MBA mind-set that demands a quick profit, a rapid promotion, and an immediate transfer. Planning for the year 2000 with five-year defense plans, three-year assignments and one-year budgets is unwise, if not irrational. As Lawrence Korb has noted, a future-oriented perspective becomes impossible to maintain when policy is shaped by short-term interests.[9] The Packard Commission suggested in 1986 that the United States needs to pay more attention to long-range planning and reward long-range planners for their efforts.

In sum, these themes may offer little optimism to those seeking to add rationality to a force planning process that at times seems out of control. But if optimism is not the message in this medium, education is. An explanation of the existence, worth, and influence of a rational process in force planning can add relevance and coherence to future debates over the size, structure, and purpose of US military forces.

2. PLANNING STRATEGIC NUCLEAR FORCES

The Background: Nuclear Strategy and Forces, 1945-1960

Although much has been written about nuclear strategies, effects, employment, weapons systems, targeting, and morality, little has been written about planning for nuclear forces. It is not difficult to see why. In American nuclear history (increasingly the only diplomatic history with which students of defense policy are familiar), strategic force planning decisions have been isolated and insulated from declaratory policies and employment doctrine.[1] Despite Bernard Brodie's prescient warnings, the atomic bomb was thought first to be just another weapon.[2] Under the prevailing doctrine in the late 1940s, in a general war against an adversary lacking nuclear capability the United States planned to launch all its nuclear weapons against the USSR as soon as hostilities began. Such an approach to nuclear weapons employment obviated a strategic force planning process that enumerated objectives, planned forces, and integrated weapons systems in a coherent way. The military support for a strategy of "massive retaliation" appears relatively insensitive to mission requirements when compared with the multiple scenarios and arcane jargon of "withholds," "second strikes," "counterforce," and "strategic reserves" that characterize our contemporary strategies. Even fifteen years into the nuclear age, an analyst seeking a rationale for the type and number of US strategic forces would have found it difficult to go beyond the declaratory and employment policy of a massive strike on the Soviet Union.

Rather, what a 1960 force planner found was a force structure developed under what some critics claimed was a strategy of "overkill."[3] Though not carefully calculated, such a force was, for its time, a rational approach to the emerging concept of deterrence. Spurred by fears of a suspected Soviet nuclear capability to deliver a devastating retaliatory blow against the United States, the American atomic strategy offered a "Sunday punch" to prevent any such retaliation. Based on this strategy, a list of military, industrial, and economic targets in the Soviet Union drove US weapons production and delivery capability. The US nuclear stockpile grew from about 1,000 bombs in 1953 to 18,000 by the end of the fifties. A US war plan in 1954 called for a simultaneous attack on the USSR by 735 strategic bombers.[4]

It was a target-rich environment and, under such a strategy, targeting doctrine led the force planner. An Air Force view in 1959 predicted 10,400 principal military targets in Russia by 1970. Covering that target array adequately, while assuring a 90 percent probability of kill, required multiple weapons planned for each desired ground zero. One force sized to accomplish target destruction featured 3,000 Minuteman intercontinental ballistic missiles, (ICBMs) 150 additional Atlas and 110 Titan ICBMs, plus a 900-plane bomber force, including not only the B-52s and B-58s already in the inventory, but the B-70 and a newly planned nuclear powered strategic bomber as well. Although not near this force level, the 1960 US strategic force was awesome for its day. Despite fears of a "missile gap" voiced in the 1960 presidential election campaign, US strategic forces were capable of implementing the strategy of massive retaliation.[5]

Few limits were placed on the planning of strategic forces until the strategy of flexible response, emerging initially from Robert McNamara's Defense Department, replaced that of massive retaliation. Alain Enthoven and Wayne Smith, the chroniclers of force planning in the period, explain that the Kennedy administration entered office concerned with two primary issues: the vulnerability of US strategic forces and the need to increase the capability of US conventional forces.[6] The effort and method to plan general purpose forces is explained in the following chapter. Here we are concerned with nuclear strategy and force planning.

Although Kennedy, McNamara, and their ''wizards of Armageddon''[7] are generally credited with the first application of rational strategic force planning, the timing is not precise. Studies by the RAND Corporation in the 1950s had already suggested the growing vulnerability of US strategic systems, and the Kennedy team based its initial strategic force decisions more on insight and intuition than on any rigorous theory.[8] Those insights—that strategic systems capable of flexible attacks were required, along with the command and control systems to facilitate such strikes—provided the basis for some ''quick fixes'' to the perceived problem. But those short-term measures would prove inadequate for long-range planning. What was lacking in 1961 was a ''theory of requirements—a conceptual framework for measuring the need and adequacy of our strategic forces.''[9] The development of that conceptual framework and its application to the planning of US strategic forces is what this chapter is all about.

The Calculations: Planning a Strategic Nuclear Deterrent, 1960-1965

Although the strategic nuclear forces of the 1950s grew without extensive planning considerations such as their final size or ultimate use, they nevertheless proved adequate to support massive retaliation. But in the 1960s, under the new strategy of flexible response, US declaratory policy changed. Secretary McNamara specifically rejected a ''full first-strike capability'' as a basis for force planning, principally because he believed that the United States was likely to suffer an unacceptable retaliatory blow.[10] Strategic forces, therefore, abandoned the traditional military roles of offense or defense. Now, they performed the deterrent missions of assured destruction and of damage limitation—respectively, a guaranteed second-strike capability to inflict an unacceptable degree of damage on any aggressor, and a capability to limit damage to US territory, primarily by destroying enemy offensive systems, if deterrence failed. According to this formulation, the assured destruction mission was the first priority, and the heart of deterrence.

Given this priority, the next step was to decide more precisely what the concept of ''assured destruction'' meant.

Force planners admitted they could not determine an exact definition by quantitative analysis alone. The final decision would have to blend judgment with analytical skills. But a requirement to meet some assured destruction criterion, however it might be defined, simplified the force planning process by introducing a force equation of finite dimensions as the effects of nuclear weapons, through an extensive testing program, became increasingly quantifiable.

Calculations to quantify assured destruction required a number of steps. First, as in earlier strategic planning, the number, type, and location of the enemy *targets* had to be determined. Second, the number, type, and yield of *weapons* required to produce a particular probability of kill were calculated. The third step was the *planning of a strategic force* to attack the targets and deliver the required weapons, while *considering the threat* to them from enemy forces. The importance of these calculations in developing and planning strategic forces requires their examination in a little more detail.[11]

Targeting—Assured Destruction and Counterforce

When we look at the nuclear issue in defense policy from a force planning perspective, the objective of deterrence tends to dominate. When deterrence became the supreme declaratory policy objective in the 1960s, the military role in support of that goal was envisioned to be a reliable, survivable second-strike force capable of assured destruction. But unlike the conventional case, the nuclear exchange being considered went beyond both empirical data and professional experience. How, then, could such a force be planned? The only way to test the stability of deterrence, it was reasoned, was to assume hypothetically that deterrence had failed. The damage sustained in the resulting attacks could then be tallied for both sides and a judgment reached about the outcome of the exchange and the extent of the destruction. If the damage suffered by the aggressor remained in the ''unacceptable'' range, then it could be deduced that deterrence would not have failed under the tested force parameters and conditions. Of course, the unarticulated assumption was that the adversary was crunching similar numbers and drawing the same rational conclusions. Unlike secret employment policy and uncertain force capabilities, declaratory policy was intended to relay this thought process to the enemy and encourage him to adopt the same reasoning.

Although such a deliberate approach to the planning of strategic forces began to bound the problem of strategic force size, planning remained rich in uncertainty. What case should be selected for testing? What objectives should be sought? Which targets should be attacked? The force planner had to consider not only what to deter, but how to deter it. Was the force designed to stop a conventional invasion of Europe? Interdict a follow-on force? Punish a homeland? Decapitate a political leadership? Different answers to these questions would lead to different proposals for different force levels.

As suggested earlier, US nuclear targeting strategies of the 1950s featured strikes against enemy military, industrial, and economic targets in a logical duplication of the strategic bombing campaigns of World War II. By the late 1950s, its targeting policy could be characterized as an "optimum mix," one facet of which was to place the atomic weapon midway between an urban area and a military target.[12] Rightly, McNamara felt that nuclear weapons could not be used in bombing campaigns as in past conventional wars and rejected such targeting schemes as unrelated to strategic objectives. But, by failing to distinguish between assured destruction and damage limitation, his policies tended to develop a volume of targets so great that the sizable forces planned to destroy them were enormously expensive.

How to quantify the concept of assured destruction emerged as a major question. During these years, one answer suggested that destruction of approximately 25 percent of the Soviet population and about 50 percent of its industry fit the criterion of unacceptable damage. Although the choice of this criterion of deterrence was less analytical than might be supposed, it did have some analytical basis. It rested more on the economic concept of diminishing returns than on some separately calculated absolute amount of desired damage. Simple mathematics revealed that after delivering the number of weapons needed for a defined level of assured destruction, additional attacks achieved little. That number proved to be 400 megaton-equivalents (EMT). Beyond the 400 EMT level lay the "flat of the curve," as shown in figure 2:1.

If the assured destruction criteria had been reached relatively easily, thereby establishing a goal for a second-strike strategic force required to wreak such destruction, discovering criteria for damage limitation, or counterforce targeting, posed greater difficulty. No doubt McNamara was attracted to counterforce policies—anyone dealing with the megatons and megadeaths of assured destruction would be encouraged to seek reinforcement of deterrent stability at some lower level of force generation. But damage limitation posed great problems for strategic force planning.[13]

McNamara retreated from his flirtation with counterforce, at least in his declared policies, for a number of reasons. First, counterforce targeting possessed first-strike implications, the very doctrine that the Kennedy administration was trying to move away from in order to achieve greater strategic stability. Secondly, it seemed unlikely that the USSR was either politically or technologically capable of adopting a policy of limited nuclear war implied by a damage-limiting strategy. Third, US allies in Europe were opposed to such a policy, envisioning Europe as a nuclear battleground for the superpowers removed from the fight. Finally, and most important from a force planning perspective, McNamara realized that a counterforce policy would generate unlimited requirements for strategic forces. Assured destruction was defined by finite levels of destruction, suggesting a strategic force limit; force levels for a counterforce policy would be limited only by the adversary's force posture. Given a Soviet buildup of nuclear forces, counterforce strikes did not offer high confidence levels of damage limitation. Preliminary force planning confirmed his worst fears. Air Force General Thomas Power, Commander in Chief of the Strategic Air Command, suggested to President Kennedy that 10,000 Minuteman missiles might be required to support such a policy.[14]

McNamara's answer to this strategy-force dilemma was to *declare* a policy of assured destruction while defining

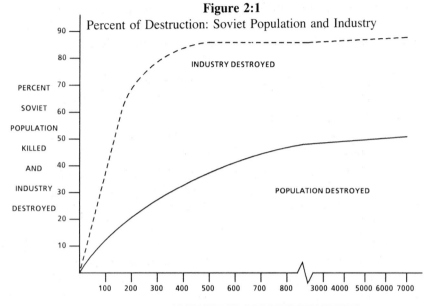

Figure 2:1

Percent of Destruction: Soviet Population and Industry

INDUSTRY DESTROYED

POPULATION DESTROYED

PERCENT SOVIET POPULATION KILLED AND INDUSTRY DESTROYED

US 1-MEGATON-EQUIVALENT WARHEADS DELIVERED

SOURCE: Fred Kaplan, *The Wizards of Armageddon* (New York: Simon and Schuster), p. 318.

PERCENT OF DESTRUCTION

MEGATONNAGE DELIVERED	POPULATION %	INDUSTRY %
100	15	59
200	21	72
400	30	76
800	39	77
1200	44	77
1600	47	77

SOURCE: Alain Enthoven and K. Wayne Smith, *How Much Is Enough?* (New York: Harper, 1962), p. 209.

employment requirements so conservatively that a flexible force capable of conducting a variety of missions would result. Thus, the priority given to assured destruction met the declared goals of stability and established upper limits for the strategic forces being planned. Furthermore, the forces generated under this strategy, it would turn out, would be numerous and capable enough to carry out missions of damage limitation as well as assured destruction.

Target Destruction

With the number of targets established, the force planner set about the task of deciding the means of destruction. To work from the declared policy of assured destruction to a proposed force level, the planner had to consider employment policy. No self-respecting nuclear operations analyst was without his well-worn Baedeker for World War III entitled *The Effects of Nuclear Weapons*. This guidebook calculated what the relationship would be between nuclear weapon yields, effects, and accuracy, and the destruction of a given target.[15] Because these relationships are spelled out in great detail in that book and elsewhere, no details of those relationships are presented here. But because calculations of strategic exchanges rely so heavily on such relationships, some key features do need illumination.

The destructive power of nuclear weapons is enormous. That's why they are treated with such deference and why they continue to form the principal security problem of our time. An atomic explosion can be broken down into the effects of blast (50 percent), thermal radiation and heat (35 percent), and nuclear radiation (15 percent). Heat and radiation will vary with conditions, and each effect will attenuate at different (quantifiable) rates according to distance from ground zero.[16] Even against hardened targets, the bomb's destructiveness is usually based on blast effect maximized by a low-altitude airburst of the weapon. A key relationship exists between weapon yield, blast effect, and accuracy. A paradox of nuclear force planning is that as the number of warheads has increased, the total number of deliverable megatons has declined. This paradox results because, as accuracy increases, yield can be decreased without losing destructive potential. So it has proven possible to gain greater destructive power through more accurate reentry vehicles rather than through larger warheads. Of course, this increasing accuracy of systems on both sides has led to fears about the vulnerability of fixed land-based ICBMs. The relationship between yield and accuracy also implies that static measures of strategic power alone, such as throw-weight or megatons, are not particularly meaningful. This relationship between yield and accuracy has led to another important index for force planning calculations: kill probability or P(k). As accuracy increases, the yield can be lowered without reducing the probability of kill of the target. For example, to obtain a constant P(k) of .95 against a 300 psi target, a range of accuracy (CEP) and yield would look like that shown in table 2:1.[17]

Table 2:1

Yield Required for Constant P(k) of .95 Against a 300 PSI Target
(1 MT = 1,000 KT)

CEP (feet)	Yield of Weapon
5,000	100 MT
1,250	17 MT
312	30 KT

This relationship allows weapons planners to reduce the yield and the weight of individual warheads and to increase their number and accuracy. We know this trend all too well as MIRVing—placing several multiple independently targetable reentry vehicles on one missile instead of having one warhead per missile. Smaller in yield, these multiple warheads can, nevertheless, cause much greater damage and complicate any enemy defense. Thus, the properties of nuclear weapons and their effects have moved both sides to acquire multi-warhead missiles of great accuracy.

Measuring Force Effectiveness

What might be called the operational level of nuclear force planning is the calculation of the effectiveness of nuclear weapons in a hypothetical exchange. Armed with a target list and a knowledge of nuclear weapons effects, a force planner can determine, under a wide range of assumptions, the outcome of a particular scenario of nuclear weapons employment. But these dynamic models of theoretical nuclear exchange, widely acknowledged to be superior to mere static comparisons of numbers of weapons or systems, are nonetheless tricky.

This trickiness results because the design of such complex scenarios requires a series of assumptions which drive the outcome. Yes, statistics can become manipulable elements even in a strategic weapons debate. Yet in designing the scenario for a hypothetical exchange, some assumptions are relatively inescapable. Which side strikes first, which weapons that side uses, and which alert posture the opponent's force is in (day-to-day or generated) are not particularly troublesome. Weapon reliability, yield, and damage expectancy have all been tested, though not under combat conditions or over real routes to the target. Operational uncertainties also play a role. Some have argued that these operational uncertainties are the dynamics of deterrence: that is, because of the risk involved, a plausible nuclear first strike cannot be briefed or believed. They might be right.[18]

But we cannot plan our forces based on a perception of deterrence or on the shadow of power. We cannot trust that a countdown to a strategic exchange will be cancelled because either the American or the Soviet leaders will blink. By constructing and comparing forces on an analytical basis, the force planner strengthens the rational input to deterrence and is not compelled to rely solely on perception or uncertainty to deter. Therefore, what is needed and what has been developed in support of the force planning process is an analytic method that supports and generates the strategic force. The method that has proved worthy over the years has been the analysis of hypothetical nuclear conflicts. Although the force is being designed to deter, it is developed from an iterative series of simulations in which strategy and forces are placed, to the extent possible, in a laboratory environment. Although these lab tests are done at a micro level, for illustrative purposes they will be grouped together.

The method of determining the effectiveness of the strategic force is based primarily on the statistical concept of expected value.[19] The key index determining the result of a hypothetical exchange is Overall Probability of Kill (OPK), also referred to as Damage Expectancy (DE). OPK is a product of both the probability that an armed warhead will arrive at the target (PA) as well as the previously mentioned P(k).[20]

The chance of arrival, in turn, depends on a series of probabilities. One of the main variables, particularly if an attack is launched without warning or alert force generation, is pre-launch survivability (PLS). Thus, while the initiator of the attack will not be concerned with PLS, survivability will vary considerably for second-strike systems—a missile in a submarine on patrol or a land-based ICBM riding out the attack. While PLS is not a concern for a first-strike system, there is nevertheless a series of events that can lower the probability of the warhead arriving at the target on time, or at all. The launch must be successful. All systems must perform as advertised in flight: the warhead must penetrate the enemy's defenses, must arm itself, must avoid other explosions in the area of the target, and must detonate at the appropriate altitude to have the proper effect.

Given this array of probabilities, even in highly dependable systems, the OPK will be statistically reduced. As an illustration, let's assume that in a first strike, 1,000 of the same warheads are to be separately launched against a given target list. The calculation of the arrival probability might look like table 2:2. If the Single Shot Probability of Kill (SSPK) was judged to be .9 (based on the accuracy of the warhead, the lethal radius of the burst, the hardness of the target and the burst altitude) then the OPK of one warhead would be .716 (.9 x 796), and the number of warheads taking effect would be .716 (1,000) = 716.

The way force planners have allowed for such degradation in the attacking force is to place two warheads on each target. The probability of the target surviving the first attack was 1 − .716 = .284, but the probability of surviving two such attacks with the same OPK is (.284) x (.284) = .08. Therefore the probability that one target will be destroyed by two warheads targeted against it given the above range of probabilities of arrival and kill is 92 percent.[21]

Table 2:2
Arrival Calculations

Activity	Probability of Success	Number of RVs Remaining
Survival	1	1,000
Launch	.95	950
Inflight Guidance	.95	902
Penetration of Defenses	.9	812
Arming & Detonation	.98	796

SOURCE: Author's calculations

This brief outline of nuclear targeting considerations, nuclear weapons effects, and levels of confidence based on probability suggests how computational techniques aid force employment and, therefore, force planning decisions. The complexity of these calculations increases greatly at the operational planning level, but the principles remain the same. Because there is no single comprehensive model of strategic nuclear war, different organizations concerned with the development of these forces and policies will use their own methodologies and measures. Although assumptions and scenarios employed in the calculations made by the military, some defense industries, the Congressional Budget Office, and the arms control community could vary considerably, that is not the point. The key point is that strategic force planning demands such methods, and, despite apparent and explicit imperfections, quantitative models of nuclear exchanges already have been widely applied to help plan US strategic forces. The following sections of this chapter suggest how.

Outcome 1: The Baseline Strategic Force

The US strategic force reached its baseline level of 1,000 Minuteman ICBMs, 54 Titan II ICBMs, and 41 fleet ballistic missile submarines with their 656 Polaris missiles in 1967.

Since that time, while modifications and modernization of that baseline force have occurred, there has not been a major change in these numbers of launchers. How were such numbers arrived at and what role did rational analysis play in determining such a force structure? More important, what can we learn from this process to aid in the planning of strategic forces for the next century?

Force planners in the Kennedy administration, desiring some yardstick of sufficiency on which to base their strategic decisions, lamented that force planning was being accomplished by the separate armed services without adequate coordination. Thus in 1961, when the Navy briefed Secretary McNamara on its need for 45 Polaris submarines, "There was not one reference to the existence of the Air Force or its weapons systems."[22] In retrospect, such an approach is not surprising. In determining strategic offensive force requirements, force planners acted with minds set on massive retaliation and without McNamara's analytical techniques. The 45 submarines were based on the calculated ability of that force to meet a finite, assured-destruction mission. But they had been built to meet this requirement without regard for land-based missiles and aircraft that could also contribute to mission accomplishment. Therefore, one might suspect that the sea-launched ballistic missile (SLBM) force could have been reduced by one-third. Why was the final number reduced only marginally to 41?

An answer to this question, perhaps the best analytical one available, is that calculations made in support of assured-destruction forces were made based on "extremely conservative" assumptions against a threat even "greater than expected" by existing intelligence estimates.[23] Thus, the guiding concept in the planning of strategic forces in the 1960s was to hedge against uncertainty by constructing *each* leg of the emerging "triad" with the ability to carry out an assured-destruction strike. Such planning clearly would result in force levels larger than those initially judged as adequate to deter.

In the case of the Polaris fleet, the final number of 41 was changed little from the original proposal of 45. Desmond Ball has argued that the limited change came about because the

Navy, unlike the Air Force, accepted a theory of "finite deterrence." Given a list of Soviet targets and the capabilities of the Polaris system, the Navy calculated that 647 missiles were needed to meet assured-destruction criteria. Forty-one Polaris boats could provide that force, and this was the number submitted by the Navy to Congress in 1957.[24]

Although the number briefly shot back up to 45 during the course of Congressional testimony over the next few years, the Navy offered little opposition when McNamara set the final buy at 41 boats in 1961. This force level represented more than just the final stopping point in a series of acquisition and production schedules; it suggested the satisfaction of the assured-destruction criteria, while also acknowledging an interface with land-based missiles. For at the same time the Navy accepted 41 Polaris subs, they also agreed to an Air Force program to procure a force of 1,200 Minuteman ICBMs. Though an optimum point or mix between the Polaris and Minuteman buys could not be determined very accurately, it was generally accepted that more than 1,500 total warheads was well past the knee of the curve and resulted in increasingly marginal returns.[25]

While Defense Department planners were striving to construct a Minuteman force that would complement the Polaris fleet, Air Force strategic planners, leaning more toward counterforce targeting policies, were not certain that a land-based missile force could be constrained so easily. Estimates of the size of the force required to meet that mission ranged from a "minimum production run" of 600 to requests as high as 10,000. Subsequent proposals suggested that the Minuteman force continued to be planned in isolation from the SLBMs and that the end strength of the program varied considerably depending on assured destruction or counterforce emphasis. In 1960 the plan rested at 1,450 missiles, but when new numbers were requested by the Kennedy administration, the Air Force ICBM proposal jumped to over 3,000, including Atlas and Titan. Although subsequent alterations to these figures were not accompanied in the public literature by official rationale, it appears that the Air Force continued to request a Minuteman force of 1,950 missiles through most of the decade.[26]

The march to the final decision of 1,000 Minutemen was an incremental one, but the choice was strongly based on the requirements of a force sized to meet assured-destruction criteria in a second strike scenario. Whatever role was played by political or bureaucratic maneuvers, or the simple attractiveness of the number "1,000," the sizing of the US missile force in the 1960s seems to rely primarily on McNamara's belief that these forces should be retaliatory in employment, synergistic in design, and assured in their destructive potential.

But because a strategy of employment had not proceeded as quickly as the declared policy—a revised Single Integrated Operational Plan (SIOP) was not completed until June 1962—McNamara was forced to separate the planning of the force from its employment. A more logical approach, and probably the Defense Secretary's original intention, was to work from force employment to force planning. Lacking either operational details of force implementation or enough time to begin a major strategic buildup, though, he separated employment planning from force planning. The result was a strategic force developed almost solely from a target list. Highly conservative planning was guaranteed with a goal of assured destruction and little knowledge of force interaction.

The strategic force that emerged was supposedly limited by a target set intended for assured destruction. But rather than dividing the mission into thirds, with each part of the triad owning a share of the required, delivered 400 EMT, McNamara chose, conservatively and with great uncertainty, that each strategic arm should be capable of accomplishing assured destruction, for a potential, deliverable total of 1,200 EMT. Thus, while forces had originally been generated based on assured destruction to avoid the unlimited demands inherent in a counterforce strategy, the forces actually developed were capable of covering almost any case of operational employment—including flexible and counterforce options. Ironically, with declaratory policy being dictated by force planning, employment policy would nevertheless be allowed to diverge from the assured-destruction strategy.

Though the outcome of the planning for Polaris and Minuteman suggests the application of a rational decision base, albeit a very conservative one, the strategic bomber force, already capable of meeting assured-destruction criteria, was not subjected to such analysis. With 600 B-52s and 1,100 medium range B-47s deployed in 1961 as the principal arm of a massive retaliation policy, (they were capable of carrying a nuclear payload greater than that of the combined ICBM and SLBM forces) the planning of additional strategic aircraft was not high on the administration's agenda. In fact, decisions were taken to limit this force through cancellation of a new wing of B-52s, an acceleration of the retirement schedules for the B-47, and cancellation of the B-70 and Skybolt (an air-launched ballistic missile) programs. These decisions were based on studies that pointed to the vulnerability of the bomber force, and a feeling within the administration that the relatively invulnerable missiles should be given priority. But bombers remained a leg of the triad capable of accomplishing the assured-destruction mission.[27]

The outcome of the strategic force planning decisions taken in the early 1960s was a baseline force composed of 1,000 Minutemen, 41 Polaris submarines, and 600 long-range bombers. That result was a blend of analysis and judgment, for the problem facing strategic planners of the time was like no other. Any theory of minimum or finite deterrence was well outside the experience of those who had planned conventional wars. Therefore, it was not surprising that the first US nuclear force would be designed to be capable of carrying out strategic bombing campaigns of the past and would be limited in theory rather than in fact.

Since the formulation of the "optimum mix" as a compromise between counterforce and countervalue targeting, US force planners have struggled with the problem of how to design a force to deter. In the early 1960s, US force planners were uncertain about the nature of a strategic nuclear war and how to design for it. Twenty-five years later this is still the case. No one strategy or planned force appears sufficient to meet the many contingencies that might apply. So forces were planned conservatively in an effort to hedge against the worst of all outcomes.

If we face such difficulties in planning forces to deter a war whose characteristics have not yet been empirically demonstrated, then we might question how we could ever reach agreement on the limitation or reduction of weapons planned so conservatively. Does strategic arms control have a future, and how might we assess the worth of a strategic arms control agreement? Analytical techniques that proved useful in the planning of strategic nuclear forces should also apply to an evaluation of proposals to reduce or limit such forces.

Outcome 2: Assessing Strategic Arms Control

Glenn A. Kent, one of the first to model strategic nuclear exchanges, inquired in 1963 into the interaction of opposing strategic forces under possible arms control agreements. In the midst of the major US effort just described to plan a strategic force for the future, Kent was concerned with the proper deployment of the systems so that each side could maintain a stabilizing second-strike capability.[28]

In this early effort, Kent noted the interaction between arms control and force planning. An arms limitation could establish an upper limit on the "kill potential" of both sides. The exchange of certain information (or an agreement not to encrypt it) or the constraining of technological innovation in strategic systems could lessen the force planning uncertainty and, thereby, enhance long-term stability.

Secondly, Kent noted that an important facet of an arms control agreement in the nuclear age would be the continued existence of an inventory on each side large enough to ensure an adequate second strike. Here he united with the force planners of the time, noting that because "unacceptable" damage could not be defined with precision, the planner was forced into very conservative estimates. Finally, Kent suggested the way toward a reduction of strategic stockpiles, arguing that an assured-destruction capability should be sought at the lowest force levels possible.

Kent's study suggested the applicability of a dynamic analysis of nuclear exchanges to the design of strategic arms control proposals. If one wishes to apply the quantitative techniques of weapons system effectiveness to evaluate an arms control agreement such as SALT or START, the methodology is the same: the evaluation of a hypothetical strategic exchange—under similar assumptions—with and without the proposed agreement. If the outcome of the hypothetical exchange after constraints imposed by the treaty appears to produce deterrence and arms race stability, then the agreement can be supported. If the treaty appears to favor either side—making the post-treaty environment less stable—the treaty should be rejected or amended. Or it may be possible, even probable judging from past efforts, that the treaty will have a marginal or negligible impact on the nuclear balance. If this is the case, the force planner is likely to be indifferent to the treaty, and arguments for its ratification will have to be advanced by other, more qualitative rationales.

The argument being presented here is that the same methodology used in the planning of strategic nuclear forces can be useful in evaluating arms control proposals. But the series of proposals and agreements intended to constrain the growth of strategic arms has not been characterized by reliance on dynamic analyses of hypothetical strategic exchanges. While this rational approach may help to decide whether to support various arms control regimes, models of nuclear war outcomes have not proved instrumental in the planning or rejection of strategic arms control agreements. There are several reasons for this. As just suggested, dynamic models of exchange calculations helped develop the force of 1,054 ICBMs—a force that remained constant from the late 1960s until the early 1980s. That force structure tended to focus strategic force planning on the static number of launchers (silos, submarine tubes, bombers) that each side possessed, rather than on the deterrent value of such a force owing to its secure second-strike capability. Thus, dynamic methods yielded static indicators of the balance. These were of little concern in an era of US strategic superiority, but problematic in a time of essential equivalence and force asymmetry. Therefore, as arms control gained prominence as a means to a stable nuclear balance, limits on strategic arms were offered in terms of force outputs (which provided a convenient focus for potential budgetary savings) rather than hypothetical exchange outcomes (which enjoyed less political support or public understanding).

Emphasis on static measures of the strategic balance in arms control negotiations also resulted from the force planning orientation of the Nixon administration. The doctrine of strategic sufficiency was driven by Congressionally imposed budgetary ceilings as well as by a growing realization that US strategic superiority was bound to be short-lived. Such a strategy accepted the quantitative criteria of assured destruction used to define strategic force requirements in the 1960s. The difference was that assured-destruction capabilities were becoming mutual. To maintain the US share of that balance, emphasis fell on the triad of strategic forces and on maintaining the numbers of forces that composed it.[29]

Reinforcing this force planning orientation was the ''participatory management'' style of Defense Secretary Laird. The departure from the OSD-centralized force planning process of the McNamara years returned autonomy to the services but inhibited a coherent plan.[30] Such an approach, coupled with defense budgets balanced among the services, encouraged a one-for-one replacement of existing strategic systems and, therefore, a continued emphasis on static measures of force structure in the arms control process.

Thus, strategic arms control has been locked in a complex process of counting various asymmetrical force properties and trying to compare them on some meaningful scale, If the process has revealed anything at all, it is that a consensus does not exist in the United States on whether arms control can play an important role in helping meet strategic force requirements. Although much of this disagreement has its roots in domestic and bureaucratic politics, there has also been a failure to appreciate the relationship between arms control and force planning as illuminated by dynamic strategic exchange calculations.

Two examples illustrate this relationship. The first examines the case of the MX. When the Carter administration elected in 1979 to begin full-scale development of a system of multiple protective shelters (MPS) for this new ICBM, it made that decision based on a number of important quantitative assumptions and factors. These included the introduction of arms control constraints on Soviet programs and the calculation of nuclear exchanges with these limited forces.[31]

As in the case of the 400 EMT criterion for assured destruction, decisions on MX basing started from the number of surviving warheads required to guarantee an adequate land-based, second-strike deterrent. A force composed of 1,000 RVs was judged adequate to put the majority of Soviet hardened, military targets at risk. With 10 RVs per MX, that meant that 100 missiles, or one-half the planned force, would have to survive.

Calculations to ensure this level of survivability and to determine the number of shelters required to protect the force worked backwards from the postulated Soviet threat. A number of assumptions were required. Only the MIRVed portion of the Soviet ICBM force would be launched against US military targets. Some amount of the force—perhaps 15 percent—would be held in reserve. The attack would aim not only at the MX system but also at other land-based ICBMs and counterforce targets. Most important to this discussion, SALT II would limit the USSR to 820 MIRVed ICBMs. These assumptions left 3,100 Soviet RVs for targeting against the MX. Strategic exchange calculations further whittled down that number. Estimates of reliability and accuracy reduced the Soviet force by approximately 25 percent, resulting in 2,300 RVs arriving over the target. Statistical probability then argued for placing 200 MXs in 4,600 shelters to achieve the desired survival of 100 MXs and 1,000 RVs.

Neither these numbers nor the real ones were cast in concrete—nor, as it turned out, was the Multiple Protective Shelter scheme. As Desmond Ball notes, the assumptions were static and the calculations highly subject to change. While the MX was a system planned on a SALT-constrained world, the arms control focus on launchers rather than warheads made the calculations less persuasive. SALT also allowed each side the development of "one new type of ICBM." For the United States, that was the MX. For the USSR, a new "heavy"

MIRVed rocket would again alter the balance. Thus, force planning can be explained based on arms control limits, but planning forces based on proposed agreements can be a hazardous undertaking. On the other hand, pre-and post-treaty hypothetical exchanges can help determine if the deterrent balance will shift in any significant fashion.

The second example of the importance of exchange calculations in arms control is illustrated by the internal debates surrounding the early Strategic Arms Reduction Talks (START) proposals in the Reagan administration.[32] Here the issue was the reconciliation of the proposal for significant arms reductions with seemingly opposing requirements to rebuild US strategic forces to support deterrence. The politicians were arguing for a decrease in the ratio of warheads to targets, attempting to fix SALT by placing controls on MIRVed systems and seeking stability through equating launchers and warheads. But the Joint Chiefs of Staff were receiving conflicting guidance for the SIOP and strategic warfighting: the number and type of targets in the USSR were being increased in US estimates. Based on their own exchange calculations, the JCS argued that a reduction in warheads on each side must go hand in hand with a reduction in "aim points"—meaning Soviet launchers. Thus, arms control negotiations and numbers were subject to domestic dispute as well.

As the strategic nuclear balance has become more complex, the relationships among force planning, targeting policy, and arms control—as suggested by these two cases—have also become more complicated. Under the concepts of assured destruction and sufficiency, it was possible to develop criteria to size and shape US strategic forces. But strategic deterrence based on warfighting or counterforce makes force planning more difficult and makes arms control and targeting policies appear as opposites in pursuit of a common goal.

Understandably then, both the Carter and Reagan administrations lost their early enthusiasm for arms control. By 1979 the first proposals of March 1977 and the hard work on SALT II had been put aside. Presidential Directive (PD)-50 relegated arms control to a position "subordinate to broader US national security interests."[33] Similarly, Ronald Reagan entered office in 1981 seeking a reformation, if not a revolution, in US arms control approaches. But by the end of his first term, evidenced by a breakdown in negotiations and a buildup in arms, that revolution was over.[34]

Calculations of hypothetical strategic exchanges used by OSD in the first case and the JCS in the second cannot solve the arms control riddle. Some, in fact, have argued that we have become too enmeshed in second- and third-order issues of the strategic balance and, consequently, ignore the real stuff of nuclear strategy.[35] Perhaps so. Yet dynamic analysis, if accorded its proper, albeit secondary, spot in strategic thinking, can help judge the balance, assess strategic arms control proposals, and plan strategic forces by focusing on outcomes and results rather than on numbers and perceptions.

What can we conclude from this application of quantitative aids to force planning in the cases examined? With regard to the baseline force, one might judge that the force, like the British Empire, was acquired in a fit of absentmindedness. Nuclear strategies of the 1950s required a bomber-intensive force to carry out air campaigns similar to World War II. And despite the restraints that McNamara and his systems analysts attempted to build into the process of strategic force planning, the objective of national security was perceived as so important in the 1960s that conservative assumptions produced a force much larger than might have been required.

If one turns to strategic arms control to limit and finally to reduce these levels of forces, the record suggests that breakthroughs are not at hand. Strategic arms control appears as much a part of the problem as a part of the solution. As Kent and Brodie suggested early in the nuclear age, strategic stability will continue to demand the continuing existence of a second-strike force. The foreseeable future appears to require a postwar reserve force, an assured-destruction second-strike capability, and forces capable of covering a range of lesser strategic contingencies. Strategic arms reductions may occur. In that event, force planning methods can help us ascertain the nature of the strategic balance and, as a result, help us judge the stability of a strategic arms reduction agreement.[36]

In addition to an enhanced understanding of the rationale behind the force structure, and as a tool with which to judge strategic stability, the greatest contribution that quantitative and dynamic hypothetical exchanges can make is an appreciation for the complexity and uncertainty of it all. We have not yet invented the war to deter—no matter how many iterations of a nuclear exchange are accomplished, the plausible strategic exchange that gains political objectives does not emerge. The collateral damage remains enormous, and no clear means or method of ending the war is evident. Answers to these larger questions of how strategic forces can be converted into some meaningful political leverage lie beyond the force planning process. However, a rational force planning process provides insights into how answers to these questions should be framed.

3. PLANNING GENERAL PURPOSE FORCES

Despite the efforts and dollars devoted to planning the strategic nuclear force, it is clear that the United States, for the majority of the post-war years, has not chosen to rely on nuclear weapons alone to preserve and enhance its national security. The first reason for this is the enormous destructive power of these weapons and the uncertainty attending their use. Related to this primary factor is a second: nuclear weapons and strategic forces have been planned for—and would be used in—the narrowest range of contingencies. Consequently, these weapons of mass destruction are not appropriate tools for dealing with most situations of international tension. Strategic nuclear forces, as the previous chapter explained, have been designed primarily for deterrence, not for their warfighting properties. And while nuclear strategies are debated and declaratory nuclear policies challenged, nuclear threats are reserved for crises of the highest order.

In an international system featuring a bipolar nuclear balance, the United States uses general purpose forces as a principal military instrument of foreign policy. These forces have been designed both to deter and to defend. The planning equations for general purpose forces become more complex as we move from the realm of the theoretical to the practical. The problem is to decide what force structure best will support the global array of US foreign policy interests. If we spend approximately 80 percent of our defense dollars on general purpose forces, and the annual defense budgets of the late-1980s were planned to approximate $300 billion, the choices are not trivial. What should we spend it on? Why?

The task should be clear. The Secretary of Defense is charged with reporting to Congress the relationship between foreign policy (US commitments abroad) and defense policy (the forces required to support these commitments). A rational basis for the planning of US general purpose forces, then, would be the creation of a direct link between the commitment and the force structure. This link has not always proved easy to construct. On one hand are forty-three defense treaties, pledging the United States to support its many, diverse, and geographically separated allies in a number of ways.[1] On the other are the general purpose forces themselves—Army and Marine Corps combat and support forces, land-based and sea-based tactical air forces, the surface and subsurface Navy, and all other conventional forces on active or reserve duty. This weaving together of foreign policy commitments, military strategies, and conventional forces has frequently resulted in a thread-bare fabric that tends to unravel.

Nevertheless, over the past two decades the belief has developed that there is a logical and rational connection—and that there should be one—between foreign commitments and force planning. To relate US foreign policy interests, the threats, and the forces needed to protect those interests, the first of two planning efforts began in the late 1940s in a National Security Memorandum known as NSC-68. But this effort, although validated by the Korean War, proved both costly and premature. It was not until 1960, under a policy conceived of as "flexible response," that the budgetary constraints would be lessened and another try made at matching foreign policy and usable force. President Kennedy's initial guidance to his Defense Secretary, Robert McNamara, was that he should not be constrained by the arbitrary budget ceilings that had forced a reliance on nuclear weapons during the 1950s. The relative relaxation of these spending limits began a process of general purpose force planning based on "what type of conflicts we anticipate, what countries we choose to assist, and to what degree these countries can defend themselves; in short, on what contingencies we prepare for."[2] With some modification, this approach has dominated the planning of US general purpose forces since the early 1960s.

Generating needs for general purpose forces based on a static comparison of numbers of men and equipment or on an attempt to "mirror image" the enemy is both expensive and inappropriate. A measure of merit for asymmetrical force postures was required to assist in conventional (and strategic) force planning. To achieve this, the Defense Department has adopted a method of planning military forces which "specifies that the way to measure the adequacy of our capabilities and to determine our programmatic needs is by analyzing hypothetical conflicts and their outcomes."[3]

How might we create an analytical model that would assist in the rational planning of conventional forces to meet military contingencies in support of US foreign policy commitments? The following considerations suggest the factors that appear to be most important in the construction of such a model.[4]

1. The first planning assumption that will drive the relationship between commitment and force planning is the *requirement*. Where is the US committed—by executive agreement, treaty or its own interest—to use military force in defense of an ally? With 43 different defense obligations to meet, force planners often choose to think of theaters of operation rather than considering the use of force only in a specific country.

2. What is the *contingency* on which we should plan our forces? How serious is such an event? How likely? What could happen? What foreign policy constraints dictate or limit the kinds of force that could be used?

3. What are the US *objectives* in each theater and contingency? Is the United States pledged to defend a specific region, government, or border? Is the regime or the territory of vital, major, or minor interest to it?

4. In each contingency evaluated, what is the enemy *threat* to US objectives? What are enemy capabilities in the region? To plan on the basis of these capabilities, given the large and expensive forces required, what are the enemy's intentions?

5. What role do *allies* play on both sides? The United States has fluctuated between planning around allies, i.e., assuming that it will have to do most of the job, and encouraging the substitution of allied manpower for US troop strength. What can the allies contribute, what does the United States need

them to contribute, and how can it stimulate that contribution? On the other side, how do the adversary's allies change the force balance?

6. How does each contingency fit in with other threats and contingencies in terms of *simultaneity of requirements*? What contingencies are related, and what is the probability of contingencies occurring in isolation? What is the priority of certain commitments as compared with others? How many wars in which theaters must the United States be prepared to fight at the same time?

7. How long a war must be planned for, and how much *warning time* can be expected? An assumption about the duration of the war will affect the stockpiles of ammunition and consumables and the structure of Army division slices, as well as resupply and reinforcement schedules and requirements. The amount of lead time for mobilization and deployment both sides expect to have available affects the level of forces on active duty rather than in reserve and the number of forces deployed forward in the anticipated area of conflict.

8. Whatever conclusions are drawn from many of these questions, they must be judged in terms of *confidence levels*. Clausewitz's concepts of "friction" and the "fog of battle" remind us that one side may benefit from the fortunes of war far more than can be determined in the planning process. But making uniformly pessimistic assumptions will drive a force posture beyond budgetary limits quickly. Is the United States willing to hedge against less probable developments in these contingencies?

This model for the planning of general purpose forces suggests the type of questions that must be asked in a rational approach to the generation of such forces. Has such an approach ever been tried? It has. And given that each postwar administration has provided somewhat different answers to these questions, the model has met with surprising consistency, if not undisputed success.

The first such effort in the post-WWII years, after rapid US demobilization was quickly followed by concern for a new adversary in central Europe, was the above-mentioned NSC-68.

Projecting a growing threat to US atomic superiority (parity with the USSR was predicted by 1954) and motivated by a need to militarize the strategy of containment to stem the advance of communism, US planners sought a suitable general purpose force structure. In the planning of this structure, the JCS applied their World War II strategic concept. They divided the globe into theaters and estimated the forces required to support American interests in each region. But these projections, which essentially called for a doubling of the remaining postwar force, gave rise to the first recorded strategy-force mismatch in the history of US force planning: the forces generated simply appeared too large for any conceivable peacetime budget. Thus the plans were scaled down to more modest but significantly enhanced levels of active and reserve forces: 27 Army and Marine Corps divisions, 408 warships, and 41 Air Force and Marine tactical fighter wings.[5]

While this plan for a general purpose force served as an administrative blueprint for the conventional force buildup that accompanied the Korean War, post-Korea policies of nuclear emphasis reduced US general purpose forces considerably. By 1960 the active forces numbered only 17 Army and Marine divisions, 376 warships, and 24 tactical fighter wings.[6] But only 11 of the Army divisions were judged combat-ready, and reserve forces were even less capable. Under the 1950s policy that emphasized a reliance on atomic weapons such a reduction was not surprising. When US strategy shifted again to conventional deterrence, emphasis was returned to the planning of general purpose forces.

Thus, the second major attempt to plan a rational conventional force posture to meet US foreign policy commitments began in the first days of the Kennedy administration. But by 1960 the United States had extended its foreign policy commitments far beyond the security borders that generated the unattainable force posture of NSC-68. The United States had signed, in what some contended were absentminded fits of ''pactomania,'' formal treaty commitments with more than 40 countries and held informal pledges to several others. The prospect of designing and generating general purpose forces to meet the threats that could occur from communist aggression or the new and terrifying "wars of national liberation" seemed daunting indeed.

In 1962 another effort was made to quantify rationally the general purpose forces to support these US interests abroad as well as to protect domestic interests at home.[7] (See Tables 3:1 and 3:2.) As suspected, the new commitments assumed by the United States as a postwar global power accelerated the need for military force in support of these commitments. Even taking into account the contributions of allies, the United States, according to this study, required a force of 55 divisions and 82 air wings to support its foreign policy.

Despite a renewed enthusiasm for growth in the defense budget, the chances of the United States raising and supporting such a force in the 1960s were no better than the hopes for NSC-68 to receive full funding in an era of massive retaliation. But these large force numbers had been generated without full consideration of the model sketched above. An application of that model would have suggested that it was improbable that the United States would have to meet all of its treaty obligations simultaneously. Indeed, it seemed even in a worst case that this country might face "only" a Soviet assault in Europe, a Chinese-supported attack in North or Southeast Asia, and a Caribbean insurgency. Understandably, the original estimate of eleven separate theaters of possible conflict was reduced to three, and what became known as the "2½ war" strategic concept, as a basis for the planning of general purpose forces, was born. The structure of the force generated under this planning exercise, as shown in table 3:2, bore a remarkable resemblance to the compromise force that had emerged from NSC-68.

Another similarity exists. The planned conventional force based on NSC-68 was reduced in the wake of the war in Korea and a move to an atomic-intensive defense policy. The "2½ war" strategy met with a similar fate when the "half war" that could have been a counter-insurgency in Vietnam expanded to take on the foreign policy and force commitments of a major contingency. Many have noted that the "2½ war" strategy was never supported in terms of force posture,[8] and, in fact, a

Table 3:1

Reconstruction of the 1962 General Purpose Force Study

Theater	Number of Divisions	Number of Tactical Fighter Wings
NATO	17	25
CENTO	4	6
SEATO	17	26
OTHERS	15	22
Strategic Reserve (CONUS)	2	3
TOTAL	55	82

SOURCE: William Kaufmann, *Planning Conventional Forces 1950-1980* (Washington: Brookings, 1982) p. 6.

reassessment of the assumptions underlying the US force planning after Vietnam resulted in the strategic concept being reduced officially from "2½" to "1½ wars."

It can be argued that this change in the strategic concept did not stem only from a strategic reappraisal, but from domestically imposed constraints of manpower, money, and public opinion that forced the Nixon administration to end the war, stop the draft, and reduce the defense budget. This suggests that force structure does not derive solely or automatically from commitments, a relationship that will be examined in greater depth in the last chapter. The formulation of the Nixon Doctrine implied that US forces would continue to be maintained for the defense of Europe as well as for a lesser contingency, placed the responsibility for providing manpower for Asian and Mid-East contingencies on US allies and reemphasized the US strategic nuclear umbrella.

There were signs that this strategic concept would also have to be adjusted to growing US concerns and commitments in other regions of the globe. The Soviet Union in 1973 implied a conventional capability that could reach beyond the confines

Table 3:2

Forces (including Reserves) Required for '2½ Wars,' 1962

Theater	Number of Divisions	Number of Tactical Fighter Wings
NATO	17	25
North or Southeast Asia	8	12
Cuba	3⅓	4
Total	28⅓	41

SOURCE: William Kaufmann, *Planning Conventional Forces 1950-1980* (Washington: Brookings, 1982) p. 7.

of Eastern Europe to the Middle East. Warning in October 1974 that deterrence of war "does not simply derive from a pile of nuclear weapons," Secretary of Defense Schlesinger called for an increase in the size of the Army from its post-Vietnam 13-division strength to 16 divisions.[9] The aftermath of the 1973 Middle East war and embargo focused attention on a new arc of crisis and the political power and potential of oil and its control. The rapidly unfolding events in Iran and Afghanistan in the fall of 1979 forced a reappraisal of the Nixon Doctrine and the formulation of a Carter policy that called for the use of force if US interests in the Persian Gulf were threatened. Once again, the time to embark upon a new program to strengthen US general purpose forces appeared at hand.

As important as the planning and allocation of forces to meet specific contingencies is to conventional force planning, the contingency method does not explain completely how force needs are determined through the analysis of "hypothetical conflicts and their outcome." In the planning of general purpose forces, how do we determine the required number of troops deployed forward in Europe? How many tactical fighter wings should be stationed with them or poised for rapid reaction to a sudden Soviet invasion? What is a rational basis on which to base the number of warships to meet the contingencies we anticipate?

To develop answers to these questions, the second section of this chapter adopts a more micro-perspective of the force planning process. Given this general model for force planning, how can we approach the actual force requirements for each contingency? The effort here will be more suggestive than complete and will use "back of the envelope" calculations rather than elegant computer models. In describing force planning for each arm of the service, I will first look at the specific mission involved, examine a method used to plan the force to undertake that mission, and finally present a specific case in which analysis can help make a force planning decision.[10]

Designing the Ground Force: The Mission

As suggested by the force planning model developed above, the design of ground forces cannot exist apart from real contingencies, cannot be based on "nice-to-have" ideals, and cannot forsake its rational basis by attempting to plan forces to meet fuzzy perceptions of military power. On the contrary, planning US ground forces requires sound and direct policy guidance. Although policy of this sort (usually contained in the classified Defense Guidance and, more publicly, in the Defense Secretary's *Annual Report to Congress*) is not well known outside force planning circles, a review of guidance to force planners as formulated since the early 1960s would probably look something like this:[11]

1. Plan forces primarily to meet the NATO contingency—an attack on central Europe by combined Warsaw Pact forces. Not only is this the most important theater for US interests, it is also the most demanding. Because a NATO contingency might develop out of a lesser conflict elsewhere, plan to be capable of defending NATO and meeting at least one other contingency simultaneously. That contingency will be less intense and less important.

2. The enemy facing the US forces in Europe and possibly in other major contingencies will be highly mobile, heavily armored, and firepower-intensive. While some warning time can normally be expected, an attack aimed at surprise cannot be ruled out.

3. Allied contributions should be assumed in Europe but will be less dependable in other regions. In NATO, members will participate in their common defense under a combined military structure. Plan on limited host nation support.

4. Reserve forces will be available to achieve the planned and required force levels but only in a major contingency and only with requisite mobilization time.

5. The conflict will begin and remain, for some period of time, non-nuclear. Conventional forces will be especially important in the first stages of the war. The USSR may preempt with nuclear weapons, or NATO may have to use theater nuclear weapons first.

6. The role of US forces in Europe, and probably in other contingencies as well, will be primarily defensive. No specific conditions for the termination of the conflict have been established. Forces should be capable of halting an initial attack and holding their positions pending reinforcement and ultimate negotiation. Plan, where appropriate, to defend as far forward as possible.

Under this kind of guidance the planning of US ground forces was oriented toward the NATO contingency and based on the following objectives: 1) US troops would be deployed along the border to avoid penetration and the loss of territory; 2) troops would be concentrated along the anticipated axes of attack as well as dispersed to counter possible flanking movements; 3) forces would be planned to maintain better than a 1:2 ratio of NATO forces to Pact forces across the entire frontier to negate a rapid Pact advance; 4) maneuverable forces were needed to avoid a Pact concentration of force and quick breakthrough and to meet Soviet second echelon forces with US and Western European reserves.

Designing the Ground Force: The Method

Armed with relatively complete, if somewhat contradictory and ambitious guidance, the ground force planner set about his task. The tools he employed in the development of such a force have been those relying on quantitative systems analysis and computer modeling. Let it be noted that, although the purpose here is to explain how a rational process can yield answers to force planning problems, no singularly competent or comprehensive model of land warfare exists. Quantitative techniques applied to the analysis of public issues have their inherent limits, and a rich literature critiques this application, particularly in defense studies.[12] As stressed in the introductory chapter, it must be remembered that rational force planning, as described here, continues to be a blending of analysis and judgment. Many of the tough issues in force planning must ultimately be resolved by a human decisionmaker. The power and promise of quantitative methods is to enhance and extend these judgments, not substitute for them.

More will be said about attempts to include subjective factors in force planning equations in the final chapter. In this chapter the intent is to describe a model of land warfare frequently used to derive forces required, based primarily on factors to which some numerical value can be ascribed. The most important of these factors have been:

1) *Firepower*—the number, variety and mix of weapons capability per combat unit;

2) *Movement*—the rate of penetration or movement across enemy lines of battle, usually based on force ratios;

3) *Warning*—the amount of time that the defending side has to deploy, reinforce, or resupply his forces engaged in battle; and,

4) *Terrain*—the existence of barriers or defended positions which can affect force movement.[13]

The brief definition of these common indices of military power used in ground force modeling and assessing a military balance raises the issue of static versus dynamic indicators. Although there is some discussion in the literature of war gaming about how the terms should be used, most of that is beyond our purpose here.[14] Suffice it to say that static measures pertain to the simple counting of opposing forces or systems such as those tabulated annually by the International Institute for Strategic Studies in their valuable document, *The Military Balance*. Within that book such categories as ''Manpower,'' ''Divisions,'' and ''Ground Force Equipment'' are counted and compared on both sides. Although a relatively straightforward measure, static indices nevertheless face problems of their own. For example, in the NATO context, the problem of which forces to count (France? Poland?) poses an issue that can significantly alter force ratios.[15] The greatest problem of this approach, which in its worst form approximates the weekly news magazine bar-chart school of force comparison, is that little attempt is made to judge how forces might perform against each other in a hypothetical conflict. In the quantity-quality arguments that frequently rage in force planning, those who advocate greater numbers of US forces will frequently resort to static indices to buttress their claims.

But it should not be assumed from this that static indicators are ''bad'' and dynamic indicators are ''good.'' Both must be considered if an accurate comparison of forces is to be accomplished. The major dimension that dynamic methods add to force planning is that of time. Thus, when two opposing forces are brought together in a simulated conflict, a dynamic method of analysis can include such variables as the time of flight of a projectile, the time to acquire a target, and the projected movement of reinforcements.[16] When coupled with statistical data of kill probability and an accurate account of firepower, a facsimile of a ground force engagement can be calculated—on a yellow legal pad or on a computer.

Whatever approach might be used to simulate a hypothetical ground conflict, the part of the model that determines force attrition on each side will drive the outcome. Factors included in the attrition equations will determine winners and losers in a single engagement, dictate the movement of the line of battle, print out the levels of manpower, equipment, and supplies that were consumed in the battle, and will thereby generate resupply and reinforcement requirements as well as the locus of the next engagement. Although the combinations of the factors employed can be varied, two approaches to attrition modeling have dominated the quantitative analysis of ground warfare.

Attrition Model I

Generally described as an index-number approach, the first model is an attempt to combine similar types of weapons in a composite index which represents the combat power of a military force, *i.e.*, its ability to inflict damage on an opposing force.[17] The most common unit of measure for this aggregation is firepower scores or firepower potential. These scores are effectiveness ratings based on empirical data and laboratory tests of various weapons and munitions. For any weapons system, such as an M-16 rifle or an AK-47, a lethal firepower score (potential) can be determined based on its ammunition, rate of fire, and accuracy. Adjustment for the type of target the weapon will be used against provides a score for it using that munition. This firepower score allows differing weapons to be compared. The ratio yielded as a result of comparing the firepower scores of the opposing armies (often discounted for reliability and hit probability to yield a lethality index) is termed the "force ratio," a dominant factor in computing attrition rates and movement of forces in the battle area. Because of the large amount of aggregation required by the index-number approach, other index number concepts were formulated.

A second dominant method of comparing force structures that differ significantly in size and weapons is known as the "weapons effectiveness index/weighted unit value." This method deals with specific classes of weapons such as tanks; defines certain weapon characteristics, such as firepower; and assigns weights based on those properties. It tends to accord a higher value to antitank and infantry units than the index-number method.[18]

At the small unit level, where the model was developed and where it appears most accurate, the approach seems unintimidating. Complexity increases, however, when the simulation includes a variety of diverse units over a wide sector.

J. A. Stockfisch uses the following example taken from the *Army Field Manual on Maneuver Control* to illustrate how the difference in firepower scores of units on the battlefield can yield advance or retreat.[19]

Based upon the range of the enemy, the type of target, and the positions of the opposing forces on the battlefield, the following firepower scores have been determined. The defender has a score of 1040, compiled as follows:

Rifle Company   540
105 mm Howitzer battalion firing in support   500
    Total   1040

The attacker has the following score:

Two infantry companies attacking frontally   1080
One tank company attacking from the flank   1200
One 155 mm Howitzer battalion firing in support   900
    Total   3180

Thus the ratio of combat power is $\frac{3180}{1040} = 3.06{:}1$.

From experience and experiment, this force ratio equates to a movement in open terrain for the superior force at a rate of 1100 meters per hour. Had the defender been in an open, rather than in a fortified position, the ratio of the combat power would have been doubled, and the appropriate movement rate is 3300 meters per hour.

Attrition Model II

The second major method of attrition modeling is the application of the Lanchester equations, a theory first developed by British mathematician (really an early aeronautical engineer) Frederick W. Lanchester. His theory of conflict, developed in 1916 to prove his point regarding the future effectiveness of aircraft in battle, describes the effects of concentration of fire on the battlefield by means of a set of differential equations. The equations have come to be known as:

Lanchester's Linear Law, representing combat where there is no concentration of force, and therefore a direct relationship exists between firepower and casualties.

Lanchester's Square Law, representing the effect of a concentration of fire—the shooters know when a certain target has been destroyed and are able to concentrate their fire on remaining targets.[20]

To illustrate the application of these laws, it may be best to use Lanchester's own example.[21] "Let us assume that one man employing a machine gun can punish a target to the same extent in a given time as sixteen riflemen. What is the number of men armed with a machine gun necessary to replace a battalion a thousand strong in the field?" The answer to this problem, Lanchester posited, depends critically on whether or not the enemy is able to concentrate his fire on a lucrative target. If he cannot, if the enemy is "searching an area at long range or volley firing at a position," then the linear law applies. The value of the individual machine gun operator becomes that of the sixteen riflemen he can replace, and the number of machine gunners required is:

$$\frac{1000}{16} = 62.5$$

Under Lanchester's square law conditions the situation is considerably different. The enemy is now able to concentrate his fire on a reduced number of targets, lessening the machine gunner's effectiveness and making his life exciting but brief. Under these conditions, the effective strength of one side is not proportional to the first power of its efficiency but to the square of the number of combatants entering the engagement. To hold its own as a battalion of 1000 riflemen (or against such an opposing force of equal firepower) the number of machine gunners really required is:

$$\sqrt{\frac{1000^2}{16}} = 250$$

So a corollary to Lanchester's square law is that it is more profitable to increase the number of participants in an engagement than it is to increase the exchange rate or the effectiveness of individual weapons. This argument that numbers can outweigh technically superior weapons is known in its current guise as the quantity-quality debate. Few realize its mathematical and historical evolution.

As a final example of a problem of quantity versus quality in Lanchester's laws, consider a tank battle on the plains of Central Europe between 60 Warsaw Pact tanks and 20 NATO tanks. Under the linear law, if the tanks are equally effective and not given the advantage of the defensive, then the 20 NATO tanks destroy 20 enemy tanks and are similarly annihilated, leaving 40 Pact tanks. If the NATO tanks are M-1s and prove to be about three times as effective as the Soviet T-54s, then the forces break even. But if the square law is applied, the NATO tanks must prove *nine* times more effective just to draw the battle. Under these simulated conditions the Pact's numerical superiority clearly presents a more serious threat.[22] Neither model is a pure fit, but these quantitative methods can aid in ground force planning. The next section suggests how.

Designing the Ground Force: US Troops in Europe

When US troops were first sent to Europe (other than as an occupation force) during the Korean War and under the mantle, if not the specific guidance of NSC-68, they were deployed as a political move rather than in response to a rational assessment of threat. The JCS had been able to establish a limit on the number of US forces that would be deployed to Korea. It was thought important to emphasize to the USSR that the United States would not be drawn into a full-scale war in Asia at the expense of the newly formed NATO alliance; indeed, there was a belief that the attack in Korea was a feint and the real attack would come in Europe.

If some sort of early ground force model were used to generate the number of US divisions required in Europe to stem the threat, perhaps along the lines of the force planning conducted under NSC-68, the number derived would have been prohibitive. Unaware of Soviet demobilization that had taken place after World War II, the United States tended to regard the Soviet Union, and the countries of the soon-to-be-formed

Warsaw Pact, as possessing enormous conventional strength. In response to this assumed force imbalance, and still acting under the premises of NSC-68 desiderata for conventional parity in an age of approaching nuclear stalemate, the NATO defense ministers, meeting in Lisbon, called for a total of 96 NATO divisions by 1954 to meet a Soviet threat estimated at 170 divisions.[23]

If Korea had validated the military premises of the strategy embedded in NSC-68, the economic and political cost of the conflict undermined the development of a large conventional force. The strategy of massive retaliation, adopted under an Eisenhower administration that had been elected on a platform of ending the war, obviated the manpower-intensive strategy of NSC-68 and reduced the NATO goals to 30 divisions by 1957. Eisenhower even flirted with the idea of reducing US forces and their dependents in Europe in order to bring down the balance-of-payments costs which threatened to unbalance the budget.

The Kennedy administration rejected such efforts. Concentrating on achieving a military balance across the spectrum of conflict, it sought to strengthen the US conventional deterrent in Europe. As systems analysts in the McNamara Pentagon set out to examine the conventional balance in Europe, the employment of quantitative and firepower models of force exchange began for the first time to take some effect. Those who had for years taken for granted the overwhelming conventional superiority of the Soviets in Europe began to wonder how the Soviet Union, with a population and a defense budget comparable to those of the United States and smaller than those of NATO's nations combined, was able to procure and support such massive ground forces while simultaneously making progress in strategic weapons and outer space.

The answers to these questions began to unfold as US and Soviet divisions were compared in terms of manpower and firepower. Static indicators of the number of divisions on each side, without regard to dynamic measures of force exchange, proved to be very misleading. In fact, it was determined that the concept of a "division" in no way described an equivalent unit of combat power in the Soviet and American armies.[24]

Further analysis indicated that a fully mobilized US division force had about three times as many people in it as did a fully manned Soviet division and cost about three times as much. Firepower scores suggested that the Soviet frontline divisions were considerably less effective than US divisions deployed to Europe; many of the Soviet divisions previously counted as combat-ready were really shells made up of cadres requiring significant mobilization and training before they could enter a conflict.

The thesis of this analytical exercise was that the United States and NATO were not hopelessly outmanned and outgunned and that there was a genuine chance for conventional deterrence to work in Europe. If such a defense were possible—and clearly, in an age of nuclear showdown, it was desirable—how could it be achieved? What forces were necessary for the United States to deploy to Europe to maintain a conventional deterrent and to hedge against the possibility of a surprise attack? How many divisions should be held in the continental United States as a strategic reserve?

Assessing the military balance in Europe is a tricky proposition, and the balance derived depends critically on the degree to which one's assumptions are optimistic or pessimistic. The purpose here is not to reach an appraisal of that balance, either in terms of how it appeared in the early 1960s or how it seems today. Rather, my point is to show how rational methods of analysis were used and can be applied, not only to assess more accurately the enemy threat, but also to derive an appropriately sized division force for deployment to Europe. Over a period of time when Senators Mansfield, Nunn, and others have threatened to withdraw US troops from Europe either for domestic political reasons or to strike a better burden-sharing bargain with our allies, it is important to realize that the number of those deployed forces has not been pulled from thin air. Rather, that force size represents a significant analytical undertaking that has determined the numbers of divisions in the US Army, both active and reserve.

The way in which the question has generally been raised is, "How many ground forces are required to be stationed in Europe to hold against a Warsaw Pact attack before reinforcements can arrive?" Most logically, and in consonance with the force planning model developed earlier, the answer to that question begins with an assessment of the threat. As systems analysts Enthoven and Smith pointed out, the 1950s canonical number of 170 opposing divisions no longer applied to the rationalization of a NATO ground force structure. With regard to enemy troops available to attack NATO's center, a relatively optimistic assumption suggested a total force of only 86 divisions (55 Soviet divisions from Eastern Europe and the Western Military District of the USSR and 31 additional Warsaw Pact divisions) available up until 14 days after mobilization, or $M + 14$.[25]

If NATO's forces are counted to include French and German Territorial Armies, NATO can field up to 26 divisions in the same time. What should be the US contribution to allow a stalwart defense of Europe? By employing a dynamic model of the hypothetical conflict suggested using the techniques outlined above, it can be shown that at $M + 4$ the force ratio derived from the firepower potential and lethality indices begins to favor the Pact.[26] By $M + 9$, with 58 divisions available to the aggressor, the growing force ratio in favor of Pact forces forecasts a NATO defeat. Although a 3:1 force ratio is often extrapolated from the work of Clausewitz and others as an advantage inherent in the defense, quantitative models have demonstrated that in a concentrated frontal attack a force ratio as low as 2:1 will gradually swing the advantage to the offense. To derive the number of US divisions that must be deployed to the theater, one has only to compute the number needed to balance the force ratio equation at each point in time. The model shows us that a deployed force of 6 US divisions (in addition to the existing allied force) is sufficient to hold the attack without considerable enemy advance through $M + 9$ against an opposing force of 58 divisions. With some variation, this is approximately the size of the force that has remained in place.

To meet the Soviet assault of 85 divisions at $M + 14$ and beyond, the problem becomes one of reinforcement rather than deployment. At this point one encounters the debate about trading territory for time. It is conceivable that the United States could save billions of defense dollars by mothballing CONUS-based divisions and airlift, relying on the deployed force to deter a conflict, and calling up reserves in the event of a major attack rather than relying on a rapid reinforcement/reaction policy. But the rate of movement for heavily armored and mechanized Soviet forces, derived from the force ratios against the deployed force, suggests that an enormous amount of territory would be lost. Thus the United States must plan to have in place a total of 15 divisions by M + 30 to stabilize the front and meet the stated political objectives. It is the requirement for this deployed and reserve force that has driven the size of the US Army and answered the ground force problem for most of the postwar years.

Certainly any overly optimistic or pessimistic assumptions can affect the size of the force, but even a range of force requirements can bound the problem; that is, provide a useful area of agreement and suggest a number of alternatives by which the same goal of firepower of force ratios can be achieved. One final example suggests this.

Consider the above case at M + 9 with some important modifications. The Warsaw Pact is still able to throw 58 divisions into the fray, but non-US NATO can now manage only 24⅓ divisions. Firepower calculations thus yield a shortfall of eight divisions, but the US has only six deployed. The quick answer to simply deploy and maintain two more US divisions to Europe, at an economic cost of some three billion dollars annually, and incalculable political cost, is unrealistic. Some have argued that the difference could made up by tactical air power. Through simple firepower potential calculations, four wings of A-10 aircraft can be shown to equate to the firepower of two divisions and to buy the time required for additional ground reinforcements. This substitution of air power for ground forces illustrates the kind of force planning solutions that can be derived through quantitative techniques but raises a larger issue as well. Is this how we should plan our tactical air forces?

Sizing Tactical Air Power: The Mission

One approach to sizing US tactical air forces argues that they should not be viewed as a separate entity in the process of force planning, but as part of the total general purpose force.[27] This means tactical air forces should be sized to fight in the same contingencies the land (and naval) forces are planned for: lending direct fire support to those ground forces, supporting them indirectly by controlling the air above the battlefield, or interdicting rear echelons/infrastructure composing an enemy follow-on force. The major portion of this section will be about tactical air support for the Army's land battle, a primary responsibility of the Air Force. The size of the Navy air support is directly related to the number of carriers it has, although land-based air can share part of that mission. The Marine air wing is sized to support the active Marine three-division force, which is fixed in number by law. These issues of naval-marine air will be referred to in the following section.

The sizing of US tactical air in support of the ground battle has concentrated, under DOD guidance, on the NATO-Warsaw Pact conflict in Central Europe. In such a conflict, as Air Force doctrinaires are quick to point out, tactical air forces enjoy unique advantages of flexibility, mobility, maneuverability, and firepower.[28] US tactical air can concentrate rapidly on Warsaw Pact attacks under a variety of circumstances, including short-warning attacks. Tactical air reinforcements, dual-based in the CONUS and Europe, can also react rapidly to the battle area. Thus US tactical air power acts as an important component (as do the allied tactical air forces) of NATO's ability to deter and defend conventionally.

In the support of the land battle in a major contingency the tactical air forces have two main purposes: 1) to defeat the enemy's air force and prevent it from interfering with allied ground operations; and 2) to attack enemy armies and installations in support of the ground forces.[29] The complex activities that current US and NATO doctrine expects the tactical air forces to perform are generally grouped into three major mission categories: counterair, close air support, and interdiction.[30]

Counterair

Also referred to as "air superiority," this mission has the primary purpose of preventing hostile aircraft from interfering with friendly ground, air, or naval forces. It incorporates the classic dogfight over the battlefield as well as air strikes across the forward line of troops to destroy enemy air assets on the ground. Advocates of the offensive counterair mission point out correctly that it is far easier to destroy enemy aircraft before they become airborne, although the task has been made more difficult with the development of hardened shelters. Deep penetration is required for such strikes, and air bases are among the most difficult of targets: heavily defended and equipped with rapid runway repair capability to reopen the airstrip within hours after the attack.

Close Air Support

Close air support provides direct air-to-ground firepower support for friendly ground forces. Here tactical air forces make their most immediate and measurable contribution to the land battle by enhancing the firepower of the army units engaged with the enemy. Proponents of CAS point to the responsiveness, accuracy, and effectiveness of munitions delivery that aircraft can bring to the surface battle. Critics of CAS contend that it is nothing more than airborne artillery with a cost disproportionate to its benefits. While the Army depends on the Air Force for its fixed-wing CAS, the Marine Corps controls its own tactical air in order to guarantee local air superiority in the "bubble" or air space over the amphibious landing area. The match of one Marine air wing to one Marine ground division has proved sufficient in the past and suggests a guide for sizing Air Force tactical air to an Army division in the close air support role.

Interdiction

Interdiction means attacking enemy ground capabilities indirectly by reducing supplies and consumables available to the battlefield, disrupting lines of communication, destroying follow-on forces, and diverting enemy resources from attack to defense. Like counterair, the interdiction mission is also characterized by an internal dichotomy depending on the locale of the target. "Battlefield" interdiction calls for attacks on enemy ground forces and infrastructure just behind the front lines of battle and is seen to have an immediate effect on the ground campaign. "Deep" interdiction calls for attacks on enemy facilities far to the rear of the battle area. Because the results of interdiction air strikes are difficult to judge, expensive to mount, and frequently misunderstood as an attempt to "strangle" the enemy army, this mission has remained the most controversial in the tactical air repertoire.

Sizing Tactical Air Power: The Method

Like his ground force planning counterpart, the tactical air force planner is constrained by the size of the baseline force, the varying missions to be accomplished, and the enemy threat to be countered. Over time, three major methods have been used to size tactical air, assuming the European scenario and using quantitative methods.

The first method, one that is incomplete from force planning perspectives, is simply the static comparison of numbers of tactical aircraft on each side. By observing the Warsaw Pact threat arrayed against NATO forces, a planner can suggest an approximately equal number of allied aircraft that must be pitted against that force. For example, at M + 1, the number of US tactical air wings that must be in place to meet the threat in a no-warning scenario can be calculated by matching the number of Pact aircraft capable of attacking NATO forces without a forward deployment. The planning exercise can then be made more dynamic by adding a time-phased deployment schedule and projecting Pact reinforcement capacity as the war continues. Balance in the initial air battle is maintained by the presence of 11 combat air wings, bringing the fighter-attack force close to parity. But by M + 14, the Pact total can be raised substantially, requiring the addition of another 1,000 NATO aircraft—translating into the rapid deployment of another 14 US tactical air wings.

Although this method can clearly drive force planning equations, it suffers from the shortcomings of other static indicators in failing to model the force balance in any dynamic way, neglecting the important relationships to the ground battle and the division of labor among the separate missions and omitting the qualitative differences between the opposing forces. Soviet Frontal Aviation, though impressive in size, includes many aircraft/interceptors designed primarily for air defense that will not directly affect the land battle. The 1:1 ratio used to compare forces also does not take into account the considerable differences in aircraft performance and capability, as well as air-crew training, skill, and flying time, widely accepted as according significant advantages to the West. (Such qualitative factors can also play a role in assessing ground force capabilities.) While Pact tactical air forces continue to gain in capability (and Afghanistan added somewhat to their air-to-ground combat experience), the reliability of Pact air forces, when removed from the ground-controlled environment characterizing their training and inserted into a free-play engagement against technically superior forces, has yet to be proved—and is difficult to simulate.[31]

A second method used to size tactical air forces is to accept the close relationship between ground and air forces and let the quantifiably determined division force drive the number of fighter wings required. If the force planning focus in Europe or any contingency remains on the number of Army divisions, as many argue it properly should, then the force planner's task is considerably simplified. A rule of thumb is that each ground division requires the support of one and one-half to two tactical air wings.[32] There is a real life guide to this planning factor— the Marine Expeditionary Force is composed of one Marine infantry division (similar in size to its Army counterpart) to one Marine air wing (which equates, with its 140 aircraft, to almost two Air Force fighter wings).

In fact, the United States has generally maintained a tactical fighter wing-to-ground combat division ratio ranging from 1.5:1 to 2:1, in part owing to the need to carry out the missions of counterair and interdiction in addition to the close air support role. If the air planner lets ground divisions drive his requirements, the six division force deployed in Europe requires the support of 9-12 wings at M + 1. But by M + 30 the 15-division force deployed needs about 23 air wings. Thus, in an Air Force composed of 26 active and 14 more reserve tactical fighter wings, more than half the wings are connected to a NATO-Warsaw Pact contingency. Other, possibly simultaneous contingencies could raise the requirement for additional in-place or rapid-reacting tactical air wings.

Although these two methods are useful to the force planner as overall guides to structuring the tactical air force, they do little to break down the force structure into mission requirements or to help determine the design characteristics and capabilities of the aircraft needed to carry out their diverse assignments. Therefore, a third method in sizing tactical air is often employed that looks more carefully at the problem of determining how many aircraft/wings to allocate to each mission. The relationship between air and ground forces cannot be forgotten during this exercise. But another way of asking the question is, "How much air power can be used to substitute for ground forces?" Although air power cannot seize territory or completely prevent enemy movement on the ground, it is clear from US combat experience that it can prove central in halting an enemy attack. How can a tactical air force planner allocate his forces among the various missions to best accomplish this end?

An answer that "it all depends" should not startle the political scientist or the force planner. While acknowledging that the specific scenario may alter the percentage of sorties assigned to each mission, the Air Force has consistently argued that air superiority must be achieved first for close air support of the ground forces to prove successful. Two basic reasons support this emphasis on at least local air superiority: 1) the air threat must be suppressed to keep attrition at acceptable levels—usually desired to be less than 3 percent; and 2) air superiority over the battlefield will protect friendly ground forces against enemy air attack.[33]

This allocation between counterair and close air support, however, must remain situation-dependent. In the event of a "bolt from the blue" conventional attack in Europe, air power can make an important contribution to halting the thrust of the initial advance. Despite relatively high attrition, the CAS mission will assume the highest priority and act as a firepower substitute for ground forces not yet in place. At the same time, the counterair mission will be reduced to air defense and local air superiority, thus concentrating on the ground attack. The question in sizing tactical air in this case becomes, "How much air power is needed to make up for firepower shortages on the ground?" The force planner, using firepower potential scores he is familiar with from the model of land warfare, welcomes this simplified solution to the sizing of tactical air support.

But the problem in this mission-area analysis is not yet solved, for while CAS sorties may be generated from firepower shortfalls, the force required to meet the interdiction mission is less identifiable. One of the problems facing the force planner is the assumption made about the war's duration. Allocating sorties to the attack of a munitions factory deep in enemy territory may have a long-term effect on the war's outcome, but it may also prove dangerous in the short term. Further, if the enemy prepositions stocks and munitions near the battle, he can negate the effectiveness of a well-planned interdiction campaign.

Recent attention centers on the mission of "deep" interdiction—the air attack on second echelon or follow-on forces—made more effective, it is argued, by emerging "smart" technologies of munitions delivery.[34] However, emphasis on such a mission will require the most sophisticated aircraft capable of long range, large payload, high accuracy, and high speed to carry it out. And while stand-off weapons may help less high-tech platforms to meet this mission, target location and acquisition remain significant problems. In any event, the type and number of aircraft, as well as the cost, to launch such a campaign will necessarily remove resources from the local air superiority or close air support battle.

History shows that the percentage of sorties devoted to each mission has, again with some modification owing to the theater or scenario, followed a discernable pattern: approximately 25 percent for close air support, 25 percent for air superiority, and 50 percent for interdiction.[35] A force planner using such a guide to construct a combat force for Europe would come close to current force planning. Based on a 23 wing tactical air force deployed, the structure might appear as:[36]

25 percent Close Air Support =	5 wings A-10
25 percent Air Superiority =	5 wings F-15
50 percent Interdiction =	10 wings F4/F-16/F-111
	+3 wings of recce/ECM
Total	23 wings

Sizing Tactical Air Power: Assessing the Threat

The planning of tactical air forces in Europe and in other contingencies must rest on the scenario anticipated—an assessment of enemy capabilities and intentions. The primary missions assigned to tactical air forces and the assets allocated to each mission will depend to some degree on the dynamics of the battle. In the case of a surprise attack, for example, the most important missions would likely be close air support/battlefield interdiction against the Pact mechanized assault in addition to NATO air defense against a Pact preemptive air strike. In the first instance, the primary targets will be the enemy ground forces; the second mission suggests a fight for air superiority over the battlefield. How can we assess the enemy threat in each case to help plan our forces?

Answers in the air-to-ground problem revolve around the questions: "How powerful are the enemy ground forces? Where are they likely to strike? And what is the firepower/movement ratio between enemy and friendly forces?" If NATO is losing, the United States may respond with increased Air Force firepower. Here, the problem of sizing tactical air support is reduced to one of comparing firepower scores.[37] The use of aircraft to slow the enemy offensive can borrow time for the ground forces to shift reinforcements from other sectors or ship them from the CONUS. The calculation of the required air force is as straightforward as the required ground force. The A-10, the primary close-air-support aircraft with anti-tank capability, can be equated in firepower scores to some part of a land division. If one A-10 sortie equals, for example, 100 rifles or a certain number of anti-tank guided missiles, and the A-10 can launch three sorties a day, then it might be demonstrated that four wings of 72 A-10s each could make up for a two-division firepower potential shortfall in NATO division strength.

Obviously, the substitution is not perfect. The principal disadvantage, as pointed out earlier, is that while aircraft can supplement ground forces, they cannot deliver a sustained volume of fire, nor can they hold or take territory. There are other limits. Sortie rates are problematic. The number of aircraft that can be "bedded-down" on European airfields is limited. Ramp space is tight, maintenance facilities are already taxed, the number of shelters for aircraft and personnel are inadequate, and airbase operability remains a significant problem. Thus the NATO infrastructure acts as a constraint on the number of CAS (or other mission capable) aircraft that can be in-place. The problem would be worsened, however, if these aircraft were dual-based (or worse, from a time standpoint, relegated to the Reserve forces); they would be late in arriving at the war.

Another limitation of relying on the A-10 as an artillery substitute is its lack of night and adverse weather capability.[38] If the Pact is able to choose its attack timing, it will likely move to minimize NATO capabilities, given that such timing does not impose even greater constraints on Soviet forces. Thus, the substitution of air for land firepower is not pure, and the force planning choice is not a simple one.

In the second case of threat assessment to be considered here, the United States is faced with countering the attacking enemy air force, an air superiority rather than a close air support mission. The central question here is "What are the plausible enemy threats to plan against?" In a book addressing that question, Joshua Epstein has designed a dynamic force planning model to measure the Soviet air threat, to bridge the planning gap between inputs and outputs, and to show how qualitative insights can be quantitatively assessed.[39] These measures of output have implications for force planning—both in meeting enemy threats and in carrying out allied air campaigns. Although Epstein enumerates the Soviet/Pact aircraft that are designated to conduct a massive "Phase I" assault of conventional air interdiction against fixed NATO targets, he evaluates that capability using quantitative data derived from US tactical air performance, making conservative estimates whenever possible.

It is this form of "worst case" planning that drives US tactical air force planning as well.[40] The effectiveness of its air attack is determined by assigning numerical values to munitions loading, numbers of aircraft, delivery accuracy, aircraft attrition, maintenance turn-around time, and sortie rates, all sent against an array of Warsaw Pact targets. In using this dynamic model of force exchange as an aid in the force planning process, no claims are made about the absolute accuracy of any number chosen to represent a particular variable. Rather, the planner must consistently ensure that any error discriminates in the opponent's favor.[41] (The implausibility of the successful accomplishment of the Pact air attack, resting on assumptions favoring the attacker, then becomes an even stronger argument.) Of course, such an approach will drive up the force needed to achieve the desired level of destruction against the enemy target list. However, the Air Force has been forced to retreat considerably from its "planning force" to a more realistic goal of 40 tactical fighter wings and, in the interim, a fiscally constrained total 35 wings. As a result, targets are left uncovered, and the risk increases.

The value of a work such as Epstein's is not only in demonstrating how military power can be measured in terms of dynamic outputs of performance rather than a comparison of static inputs but also in a wider application to the force planning process. If Epstein's work has suggested that one of the most feared elements of the Soviet blitzkrieg—the initial air interdiction attack—is implausible, then further questions arise: What are the plausible threats, and how should we plan against them?

Similar approaches can be applied to each theater or contingency in a dynamic version of the general purpose planning model already constructed. Epstein reminds us to regard with suspicion force planning based on some general and ill-defined Soviet threat and instead to plan against a specific threat that can be realistically assessed in a specific region. Yet, this distinction between a "global threat or global war" and a set of well-planned-for contingencies is frequently not made—with serious implications for a rational force planning process. The planning of the US Navy provides us with a good example.

Shaping a Navy: The Mission

In planning for a general purpose Navy, two missions, and the relative importance granted to each, will drive the force requirements.[42] The first of these missions is *sea control*: the ability to keep the sea-lanes open to friendly shipping. The second is *power projection*: the capability of exerting sea-based military force against enemy objectives on the shore. Again, the most demanding scenario against which these forces have been planned is that of US participation in a NATO-Warsaw Pact conflict in Central Europe. (Southwest Asia is another rigorous test and may demand much of the Navy, depending on the adversary.) But while this contingency poses the most rigorous case for planning a navy, the force required to meet such a challenge does not emerge as a clear-cut choice, for the line between the two primary missions is somewhat blurred. The US Navy has generally accorded highest priority to the sea control mission, but it has not agreed that the mission can best be accomplished by emphasizing defensive operations and systems. Offensive operations against the threat to the shipping lanes offer another alternative—surely a more adventurous and, perhaps, costly one—to a concentration on the defense of sea-going convoys.

A second reason that force planning for the Navy is problematic when related to specific missions is that similar types of ships are required for each mission. The key offensive grouping of the Navy, the carrier task force, could certainly perform the sea control mission during a major conflict in the North Atlantic. But the carrier, with its tactical air arm, is also the primary naval instrument to assault Soviet assets in their home waters or on their home territory.

Further, because the carrier battle group dominates the naval force structure owing to its dual capability of sea control and power projection, an emphasis on one mission or the other can significantly affect the total number of ships required. Greater reliance on taking the attack to the enemy, for example, would require more carriers (as well as submarines in the "direct support" mission and perhaps surface ships in the battleship-based "surface action groups"). Requirements for the protection of the sea lines of communication (SLOCs) will vary with the number of convoys or units to be protected, but would likely emphasize a greater number of surface ships as well as land-based patrol, anti-submarine, and air defense aircraft, and may not need as many carriers. The fundamental issue in naval force planning, then, is whether the United States wishes to emphasize the offensive power projection or the defensive sea control mission. This choice of strategy will act to drive the number of combatant ships in the Navy.

As in the planning of other forces, the priority granted to each mission will also be a function of the assumptions about the type of war likely to develop. The Navy wishes to strike the opposing force quickly, but marshalling the power projection fleet will take time. The sea control mission becomes increasingly important as the war drags on, for only sealift is capable of transporting the sizable forces needed to reinforce and resupply a land war in Europe. Successful conclusion of a "long war" will therefore depend critically on accomplishing the sea control mission— by whatever means.

There are other missions as well. An amphibious assault with Marine forces is usually closely associated with the naval mission of power projection, and for force planning purposes is subsumed under it. Another important naval mission is that of "presence" or "showing the flag."[43] That mission has little utility for the force planner because a solid premise for sizing the force for deterrence or warfighting is lacking. Presence, therefore, has not been widely regarded as a mission capable of driving the force structure. Moreover, the types of ships generated for the primary wartime missions should be capable of accomplishing the predominantly peacetime mission of presence.

Although the kinds of ships necessary to demonstrate US interest in Third World waters need not be sophisticated systems designed for survival in a high threat environment, sustained operations in distant regions could generate requirements for more noncombatant ships, depending on their remoteness and the availability of nearby bases. Recently assumed US commitments in the Persian Gulf and Indian Ocean have produced a variety of measures (maritime prepositioning, new bases, facilities and access agreements) to support operations in limited contingencies. This topic will be treated more completely in the following chapter.

The salient point is that divergent missions lead to varying requirements in total numbers and capabilities of ships. Before examining a case study of the 600-ship Navy, however, it is necessary to point out how, given a specific task and mission, the number of ships required is generated.

Shaping a Navy: The Method

Whatever decisions may be reached about the priority accorded to the naval mission, the number of ships generated by those requirements tends to emerge in force "packages" that result from the task force concept. The aircraft carrier, the most flexible naval system for mission capability, drives the naval force planning process. But because the carrier has primarily been seen as offensive, the need was created to defend this lucrative target from enemy attacks.[44]

The task force concept, a number of destroyers and cruisers escorting the ship on its offensive mission, was judged to be the best answer to aircraft carrier defense. Over time these escort ships have become specialists in defense and have been designated as "anti-submarine" or "anti-air" destroyers and cruisers. The carrier battle group has especially important implications for naval force planning. The number of carriers afloat will be the central determinant of the size of the entire fleet because each carrier requires a certain number of escorts. Both carriers and escorts in the task force also require replenishment and support ships which, in turn, require protection. Thus the addition of one aircraft carrier to the fleet generates requirements, depending on the mission, for an additional 17 or so ships, not including auxiliary or overhaul requirements. A typical force generation for a single carrier resembles the following:[45]

CV	1	(Aircraft Carrier)
UNREP	4	(Underway Replenishment)
Aegis	1-2	(Cruisers or Destroyers)
CG	6	(Guided Missile Cruiser)
SSN	1	(Attack Submarine)
FFG	4	(Guided Missile Frigate)
	17-18 Total	

While the power projection mission dominated by the carrier task force drives the number of ships in the fleet, other naval missions utilizing the task force concept will add to the total numbers. For example, the problem of defending the sea lines of communication will be extremely sensitive to the number of units in the convoy that must be protected. And the more convoys required for ocean transit—in the Atlantic SLOC conceivably as high as five convoys of 100 ships per month—the more surface ships, submarines, and patrol aircraft will be needed. The convoy escorts, however, can be considerably less offensively armed than ships intended to assault enemy shore positions if an effective barrier of land-based aircraft, carrier-based air, and attack submarines is established along key Soviet access routes to the sea lanes. Then, the convoy escorts will need to deal only with the residual air and sea threats which repeatedly penetrate this barrier.

A typical force generation package for the SLOC mission might appear as:

For the submarine barrier:

Forward patrol sub	5
Barrier patrol sub	25
Total	30

For the surface combatants (per convoy):

Destroyer	1
Frigate	9
Total	10

Aircraft carriers might be a part of the SLOC package as well, so the cumulative force for projected control of the Pacific sea lanes, for example, including two carrier task forces, a submarine barrier, and enough escorts for two convoys per month, would equate to approximately 95 ships.[46]

This force planning exercise is a relatively simple one; the methods are straightforward. What will drive the number of ships in the fleet is not disagreement over the size of the task force, but rather answers to larger strategic questions. Is the Navy's role primarily a defensive one of preserving the security of the sea lines of communication, or should it emphasize the offensive mission of power projection to sweep the enemy from the seas? The answer to this question will size the force, as the following case study illustrates.

Shaping a Navy: Why 600 Ships?

In 1976 the National Security Council proposed a fleet sized to blend the missions and the force requirements of sea control and power projection.[47] The program, calling for a limited purchase of large, multi-purpose warships capable of the power projection mission, emphasized sea control. The argument was that the Navy, with few changes in its then existing structure, was capable of meeting both missions. This alternative fleet would be based on a 12-carrier force, with the number of ships in the fleet totaling just over 500.

A 500-ship Navy is one of considerable force and flexibility; the rationale offered for such a fleet appeared sound. But naval strategists assuming responsibility in 1981, faced with growing Soviet naval capabilities, rejected such a fleet, on the grounds that a force that size was incapable of supporting US commitments and meeting all contingencies. The Navy, now with administration support, proposed again a buildup to 600 ships.

The "600-ship Navy" has become a commonly used buzzword—if not a *cause célèbre*—for the architects of a new US maritime strategy. But like other static indicators, the number of ships reveals nothing of force capability, nor does it specify how these forces would be used in specific contingencies. Applying the framework of missions and forces developed previously, however, allows us to consider the strategic and analytical basis on which this call for a 600-ship Navy rests.

When an alternative fleet of 500 ships is compared with a 600-ship Navy in a contingency analysis, the major differences in the size of the forces stem from different priorities granted the mission of sea control and power projection, as well as different perceptions of the threat. Let us assume that agreement exists between the two fleet formulations about the ships required to accomplish the naval missions in the Mediterranean, the Pacific SLOC, and the Persian Gulf. In the Atlantic, however, a disagreement in fundamental maritime strategy equates to a substantial divergence in the number of ships required to implement each mission.

One difference in naval strategy and forces centers on the Murmansk/Norway contingency. The naval objective is to deny the Soviet fleet access to allied Atlantic SLOCs. The "maritime strategy" plans to assault Soviet shore bases in a face-to-face shootout: power projection in the extreme.[48] The alternative 500-ship fleet proposes a smaller number of ships to contain and reduce by attrition Soviet forces by emphasizing the sea control mission.

The difference between the two fleets in planning for this contingency is based on two major factors: the size of the carrier force required and the size of the amphibious force desired. The Navy proposal to carry the battle to the Kola Peninsula will require at least a four-carrier task force: two carriers to support the amphibious invasion and two to counter Soviet air power.[49] The purpose of this maritime offensive would be to destroy the enemy sea and air forces prior to their attempt to break out of the Murmansk area, invade Norway, and penetrate or bypass the Greenland-Iceland-United Kingdom gap (GIUK). This larger fleet also includes an amphibious force of three brigades to neutralize Soviet air, surface, or subsurface forces in the North Atlantic.

Planning that emphasizes sea control regards the above approach as highly dangerous and, more to the point, unnecessary. Such a shoot-out may require a force of six or more carriers rather than four to assure success. But even if the shore attack proves successful, the probability of the survival of Soviet submarines remains very high. Therefore, an attack on Murmansk does not obviate the need for a barrier to protect the Atlantic SLOC.[50]

Regardless of the strategy employed in the north, then, it is still necessary to defend Atlantic sea lanes. This second area of significant disagreement betweeen the two programs is a determining factor in the ultimate force posture. Although room remains for strategic choice, the issue is more concrete: How can the United States best defend the sea lines of communication against the Soviet submarine and naval aviation (Backfire) threat?

Solution of the problem is critical. The resupply effort will depend upon the course of the war in Europe. The longer the war continues, the more vital sea control will be to the US and its allies: 95 percent of the tonnage required—estimated at five million tons per month—must go by sea. Thus, a requirement of five convoys per month, the attendant escorts, and a submarine barrier are constants agreed upon in the strategies of both force proposals.[51]

Table 3:3

Aggregate USN Program by Specific Contingency/Number of Ships
(600-ship Navy)

Contingency	CV	Aegis	CG	DD	FFG	Amp	URG	SSN	MCM	Aux	TOTAL
Murmansk	4	8	24	9	25	51	16	4	9	15	165
Thrace	2	4	12	—	8	—	8	2	—	4	40
Persian Gulf base denial	2	4	12	6	14	34	8	2	6	9	97
Atlantic SLOC	4	8	24	5	61	—	16	34	—	15	167
Pacific SLOC	2	4	12	2	26	—	8	32	—	9	95
TOTALS	14	28	84	22	134	85	56	74	15	52	564
Overhaul	3	5	13	4	21	—	—	19	65	—	—
TOTAL	17	33	97	26	155	85	56	93	15	52	629

Plus 25 fleet ballistic nuclear submarines (SSBNs), yielding a force of 654

SOURCE: Author's estimates

This debate is by now the familiar one that drives the naval force structure—the number of carriers required to provide adequate air cover for the convoys. Each fleet is likely to be satisfied with four barriers to halt the penetration of Soviet anti-shipping forces: the submarines on patrol, air cover (including land-based P-3C ASW aircraft as well as carrier-based F-14s), mine fields, and close-in carrier/convoy defense. But a 600-ship fleet will require four carriers; the smaller fleet attempts to accomplish defense of the SLOC with two.

In contention here is the composition of the air barrier in the GIUK. Rejecting a passive, defensive strategy centered on the GIUK gap, the Navy attempts to construct the barrier with carriers, arguing that four are required to provide overlapping coverage.[52] The sea control fleet sees these two extra carriers as unnecessary and argues that they can be replaced with land-based

Table 3:4

Aggregate Alternative Program by Specific
Contingency/Number of Ships (500-ship Navy)

Contingency	CV	Aegis	CG	DD	FFG	Amp	URG	SSN	CMC	Aux	Tot.
Murmansk	2	2	12	6	14	34	8	2	6	9	95
Thrace	2	2	12	—	8	—	8	2	—	3	37
Persian Gulf											
base denial	2	2	12	6	14	34	8	2	6	9	95
Atlantic											
SLOC	2	2	12	5	53	—	8	32	—	11	125
Pacific											
SLOC	2	2	12	2	26	—	8	32	—	8	92
TOTALS	10	10	60	19	115	68	40	70	12	40	444
Overhaul	2	2	9	3	18	—	—	18	52	—	—
TOTAL	12	12	69	22	133	68	40	88	12	40	496

Plus 25 fleet ballistic nuclear submarines (SSBNs), yielding a force of 521

SOURCE: Author's estimates

air in the GIUK. Tables 3:3 and 3:4 show how the difference in assumptions—and the resulting demand for carriers—drives the total numbers in the fleet.

The maintenance of the large existing conventional land, sea, and air forces is justified by requirements to meet US foreign policy commitments. But the matching of forces to meet these commitments has become a complex process, and dealing with these important but often emotionally laden issues of budgetary import at times obscures rational approaches to force planning options. Nevertheless, a logical structure for force planning is essential, does exist, and has proved worthy of the effort required. A framework exists for rational choice in conventional force planning.

The contingency-based approach to force planning was derived in part from the development of a US strategic concept—a statement of how many and what kinds of wars the United States should be prepared to fight. A realistic construction of such a concept allowed the United States in the postwar years to plan forces based on first a "2½ war" and then a "1½ war" strategic concept. The one major contingency that has survived some 40 years has been the possible NATO-Warsaw Pact conflict in Europe, and that contingency has largely dictated the size and composition of US general purpose forces.

But what of the lesser, non-Soviet contingency, the so-called "half war?" How has the United States set about planning rapidly deployable forces to meet that threat? The next chapter explores answers to that question.

4. PLANNING RAPIDLY DEPLOYABLE FORCES

The Duke of Wellington once remarked that "a great nation cannot fight in a small war." That aphorism could well be applied to the United States since its emergence as a great power after World War II. Certainly, it encountered its share of political and military difficulties in attempting to constrain its involvement in the limited conflicts in Korea and Vietnam. This is not to suggest, however, that the United States had failed to plan forces for a war that might be limited in scope, tactics, region, and objectives. On the contrary, as suggested in the previous chapter on general purpose force planning, the need for conventional forces to implement the policy of containment across a spectrum of potential conflict had been foreseen before Korea. By 1955, with NSC-68 validated if not supported, it was accepted that US armed forces, in addition to having the capability to meet major contingencies, should include "forces strong enough to deter or to suppress small-scale aggressions or disorders inimical to American interests in the 'gray areas' of the world."[1] By 1979, American policymakers concluded that the following objectives should guide force planning for the limited contingency or "half war":

- The United States should be able to protect critical alliance interests that are endangered by a non-nuclear attack on the periphery by meeting such an attack at its own level;
- the US response should be rapid enough to frustrate a quick takeover;
- this capability should not detract from the United States' ability to fight or deter a large war happening simultaneously or sequentially;
- the United States should have a reliable capability that will meet a high level of confidence.[2]

However, as the events in Afghanistan and Iran unfolded in late 1979, and as US military responses to these events were evaluated, it became clear that US conventional forces were not adequately designed, organized, or supported to counter a modern army on a Third World battlefield. This chapter focuses on past efforts to develop a military strategy, construct a coherent organization, and procure adequate mobility systems to create a limited contingency force to meet threats to American and allied interests in a less-than-major conflict.[3]

The force planner determines how many units of various systems should be procured and operated to carry out a given strategy under a wide set of circumstances. Force planners considering a limited contingency over the last 20 years have faced great uncertainty. Vague references to US global interests prove to be insufficient guidelines on which to base a purposeful military strategy or a cohesive force structure. Thus, conventional force planners for non-specific limited contingencies were provided only general objectives. To hedge against uncertainty, they attempted to construct a flexible force for operation in an unknown environment. These efforts elevated concepts of organizational flexibility and force versatility to the *sine qua non* to meet a limited contingency. Without specific objectives, a number of possible scenarios resulted in a range of military objectives that required a variety of armed forces. Under these ill-defined conditions, plans remained unclear, deployment schedules uncertain, and the consequences of simultaneous contingencies unforeseen.

In the past, because a limited war was, by definition, of less importance than a major military effort (one definition of a limited war is a war you can afford to lose) and a lesser threat to US interests, it was tempting for the force planner to think that the military capability to handle one or two major contingencies implied the ability to deal with a limited one. Thus, while the limited contingency or "half war" has been an element contained within the US strategic concept for the last two decades (as in the "2½ war" and "1½ war" policies), forces to meet a limited contingency were never adequately planned for, funded, constructed, or supported owing to deficiencies in the following areas:

- The *strategic concept* that depended on the flexibility of a central strategic reserve was unrealistic in terms of available resources and over-optimistic about ground unit versatility.

- The *organizations* structured in support of that strategy encountered problems of unified command and faced interservice conflict over assigned missions. Forces assigned to these organizations were not dedicated to the limited contingency and were employed in combat situations by other command organizations.

- The *mobility systems* in support of a strategy of rapid deployment were not procured in numbers adequate to support a limited contingency simultaneously with a major contingency.

The following discussion suggests the planning shortfalls in each of these areas that prevented the United States from constructing a coherent limited contingency force.

Rapid Deployment Strategic Concepts

The strategic concept—the statement of how many and what kinds of wars the United States should be prepared to fight—is not a complete or precise basis for force planning. As an element of that strategic concept from 1960 to 1980, the "half war" planning contingency took on a variety of meanings, locales, and levels of intensity. Thus, the "half war" was envisioned as a limited contingency in the Western Hemisphere (early in the Kennedy administration), a counter-insurgency in Vietnam (early in the Johnson administration), a non-Asian contingency "elsewhere" (early in the Nixon administration), and a limited contingency in the Middle East (early in the Carter administration). (See tables 4:1 and 4:2).

Regardless of the strategic concept in which it was embedded, the "half war" as a basis for force planning has been rife with uncertainty; it derived from a complex and changing threat, and the belief that the United States, as a global power, must have flexible forces capable of meeting worldwide interests. Until the "half war" was assigned a specific scenario, a particular threat or adversary, and an assigned region of

Table 4:1

The Allocation and Deployment of Major General Purpose Forces
Under the '2½ War' Strategic Concept, 1965

		Contingencies		
1 War	**and**	**2 War**	**+**	**½ War**
Warsaw Pact attack in Europe		Chinese attack in Asia		"Brushfire" in Western Hemisphere
Allied Support		Limited Allied support		Some Allied support
Vital to US		Vital to US		Not vital to US
		Forces (Army and Marine Divisions)		
CONUS	4	3		—
Deployed	5	4		—
Strategic Reserve	—	—		3⅓
Reserve	8	—		1
		Strategic Concept		
Duration of conventional war in Europe limited to 3 months		Did not allow for simultaneous attack in Korea and Vietnam		Rapid deployment to come from strategic reserves
Forces in place adequate to meet intermediate-level attack		Holding action required while Reserves mobilize		Forces allocated based on simultaneous contingencies
Reserve mobilization required		Reserve reinforcement required		Strategic reserves could also be used for major contingency reinforcement

SOURCES: Enthoven and Smith, *How Much is Enough?* p. 215; Charles Schultze, ed., *Setting National Priorities: 1973 Budget* (Washington: Brookings, 1972); and author's estimates.

Table 4:2

The Allocation and Deployment of Major General Purpose Forces
Under the '1½ War' Strategic Concept, 1973

Contingencies		
1 War or	**1 War** +	**½ War**
Warsaw Pact attack in Europe	Chinese conventional attack in Asia	Lesser contingency elsewhere, possibly the Middle East
Allied support	Limited allied support	Allied support questionable
Vital to US	Vital to US	Not vital to US

Forces
(Army and Marine Divisions)

CONUS	4⅔	2⅓	—
Deployed	4⅓	1⅔	—
Strategic Reserve	—	—	3
Reserve	8	—	1

Strategic Concept

Nuclear capability of U.S. strategic/theater forces serves as a deterrent to full-scale Soviet attack in Asia

Prospects for a coordinated 2-front attack on U.S. allies are low because of the risks of nuclear war and the improbability of Sino-Soviet cooperation

Reserves may not have to mobilize

Non-Chinese threat in Asia, Middle East, or Latin America; force planning dependent on allied contribution

In case of subversion, guerilla war, wars of national liberation, U.S. will not be involved, will preempt through economic development, social reform

SOURCES: Kissinger, *White House Years*, p. 220 ff; Gelb and Kuzmack, "General Purpose Forces," p. 208; and author's estimates.

responsibility, a force planner looking at a limited contingency was required to support operational plans that spanned the globe, to design forces capable of responding to a range of conventional conflicts, and to consider a wide array of enemy threats.

Thus, the "half war" as a guide to force planning for a limited contingency attempted to embrace a wide range of conventional conflict under a single convenient, but imprecise, concept which led to disparate conventional scenarios melded into a single threat that could be countered with a single general purpose force. This aggregation has produced two significant trends that have affected force planning over the last 20 years. First, rapid deployment forces designed to fight this "half war" were conceived as flexible and sophisticated organizations composed of versatile forces that could meet any military challenge. Secondly, the requirement for the deployment of this force simultaneously with the reinforcement of US forces in a major contingency called for separate strategic mobility systems. However, although additional forces were allowed for and allocated to meet the "half war" in whatever form, these units were never provided the strategic lift to guarantee rapid deployment.

Unfortunately, these two trends were reinforced, rather than questioned, in the wake of the American withdrawal from Vietnam. Although the perception of Vietnam as a "half war" led to the initial US involvement there, the uncertainty of US objectives and the attendant escalation of that conflict resulted in a military effort worthy of a major contingency but a political commitment far less than that. Therefore, although a reflection on the United States' experience in Vietnam now appears to demand that future limited contingencies be more precisely enumerated in terms of locale, adversary, level of intensity, and allocation of forces, any earlier attempt to delineate the "half war" in more precise terms was blurred by a postwar retrenchment in both strategy and force posture. In attempting to steer a course between overinvolvement and underinvolvement, the Nixon Doctrine practically ruled out future US land conflict with Third World armies.

The perception that the United States had ineffectively wielded the military instrument in support of its diplomacy in the Third World extended into the Carter administration. Secretary of Defense Brown, in accepting the ''1½ war'' strategic concept, retreated from ''half war'' planning, acknowledging that, while the flexibility inherent in a system of strategic mobility and a ready reserve force should be maintained, ''we can probably make a larger proportion of our ground and tactical air forces more specifically equipped for operations in Europe than in the past.''[4]

However, the threat suggested by the extension of Soviet interests and control to areas well beyond central European borders in December 1979 caused that emphasis on the NATO contingency to be reevaluated. While an emerging power vacuum in Iran demanded US presence, Afghanistan was interpreted as a completely new species of provocation. For the first time since World War II, Soviet rather than surrogate forces were used to maintain Russian control in a region generally regarded as lying outside the accepted sphere of Soviet influence. With the Soviet invasion, the focus of US strategic concern shifted to Southwest Asia; the capability to meet Soviet forces on a Third World battlefield was raised to a high priority.

Subsequent examination of additional limited contingencies the United States might face revealed a host of additional threats that a Eurocentric strategic concept had overlooked. In addition to simultaneous conflicts in the Persian Gulf and NATO Europe, the United States could also be required, in the event of a North Korean attack on the South, to deploy reinforcements to Northeast Asia as well. US defense officials, in admitting that American capability to respond to ''1½ wars'' was less than adequate, conceded that US strategic mobility support and fighting forces would be stressed to the point of failure by this ''one and two half-wars'' scenario.[5]

In 1980, then, the probability that the United States might have to face more than one limited contingency simultaneously appeared to be increasing. Thus, one ''half war'' as an element of the strategic concept and as a basis for force planning no longer retained its relevance: the likelihood of sequential or simultaneous limited contingencies required forces to meet each event. These would now have to be planned and allocated on a priority system that took into account US commitments and interests, and they would have to be programmed within realistic budgetary constraints. For effective and enduring force planning for lesser contingencies, it appeared necessary to reject past concepts of flexible organizations composed of versatile forces supported with as-available strategic mobility systems. A revised strategic concept suggested the need to be prepared for a series of specific, significant, sequential, or possibly simultaneous contingencies.

Rapid Deployment Organizations

The uncertainty attending the strategic concept of rapid deployment to an undefined limited contingency made more difficult the organizational task of uniting available forces to accomplish this mission. From this uncertainty grew the requirement for a large and complex organization malleable enough to meet these ill-defined goals. But large organizations, particularly those imposed over already existing bureaucratic structures, are not inherently flexible. Earlier attempts to organize a coherent limited contingency force revealed flaws in the deployment and employment of assigned forces to meet all circumstances and contingencies. Moreover, the atmosphere surrounding the organizational planning for a limited contingency force has been unfavorable. Torn by service parochialism, deprived of assigned forces and denied budgetary support, these organizations failed to meet the limited contingency mission.

For example, the Strike Command was formed in 1961 to place a higher priority on a conventional response to limited wars. However, it was hampered from the start by conflicting organizational interests, lack of a joint service doctrine for combined operations, and the Navy's reluctance to participate. Even as a joint command restricted to ground and air forces, it was unable to resolve the doctrinal dispute over Air Force close air support and Army air mobility.

Such disagreement crippled the command. Because force employment was likely to occur in an area with little established infrastructure or command organization in place, it was

imperative for the services to develop procedures of close cooperation and coordination. Without a specific area of responsibility or an assigned mission, joint doctrine succumbed to
more powerful service interests.

Strike Command could have accomplished a limited uni-
fication of the Army's STRAC (Strategic Army Corps) and the
Air Force's CASF (Composite Air Strike Force), but these
forces, allocated to a strategic reserve rather than assigned to
STRIKE, were soon absorbed by the Vietnam war. Thus, Strike
Command, instead of becoming a unified combatant command,
merely facilitated the deployment of US forces to Southeast
Asia. Even the assignment of a geographic area of responsibility
in the Middle East proved futile, for forces deployed to that
region under STRIKE were employed by another unified command. Lacking organic naval forces and deprived of its ground
and air forces by an unforeseen major contingency in Southeast
Asia, Strike Command by 1968 possessed few combat-ready
forces to deploy to any limited contingency.

In addition to stripping Strike Command of its operational
capability, the Vietnam war also encouraged a widespread public aversion to any expeditionary force. Forces designed to
deploy rapidly to global trouble spots were now seen as stepping
stones to wider, and unwanted, US commitments. In a time of
retrenchment and reconsideration of US global military involvement, Strike Command was at best anachronistic, at worst
provocative. If, as the Nixon Doctrine implied, the United
States would no longer provide the manpower to fight limited
wars, an organization designed to deploy such forces appeared
as a nonessential luxury in a defense budget noted for its austerity.

The transformation from Strike Command into Readiness
Command in 1969 also reflected a change in the perception of
means available to conduct limited contingency operations as
well as an alteration in the perception of the post-Vietnam
threat. A growing isolationist sentiment, plus the passage of legislation limiting US involvement in Southeast Asia (as well as
the War Powers Act), suggested that budgetary support for rapid
deployment forces would be curtailed. Readiness Command was

directed to augment oversea commands with US-based forces, but was not assigned an area of responsibility, nor was it seen as the parent command of a limited contingency force. US ground forces would fight on the plains of central Europe, only. As a result, force planning for the "half war" came to a halt in the early 1970s.

As a new US administration took office in 1977, the campaign promises of the incoming President and the limits set by the public mood continued to restrict US military pledges of assistance solely to the major contingency. Within the Carter administration the increased interest of some senior officials in extending United States intervention forces occurred at a time when its capability to act was constrained by public opinion and by the administration's reluctance to request additional funding for such a capability. Although Secretary Brown alluded to the development of a strike force of highly mobile units that could be deployed to Southwest Asia, the makeup of these units, the potential adversary, and the organizational framework for the force remained unspecified.

The 1979 armed entry of the USSR into Afghanistan suggested that US forces might encounter Soviet armies beyond the confines of a NATO scenario. But the Soviet action also brought with it old fears of a global communist challenge and the need for the United States, once again, to be able to respond with military force to any global contingency. This wider view of US/USSR competition was a major departure from the post-Vietnam US defense policy extended from the Nixon/Kissinger Doctrine which tended "to consider third world crises in their specific local and regional setting, rather than to press them into the matrix of global East-West competition."[6] President Carter's personal reappraisal of the seriousness of the Soviet threat suggested the need to respond to this threat with conventional forces and strengthened the requirement for a counter-intervention capability.

This heightened perception of the threat, coupled with uncertainties about the utility of military force in the region of the Persian Gulf, resulted in a Rapid Deployment Force that carried with it the organizational baggage—and military impotence—of past failed efforts. The first uncertainty regarded the adversary. Should the force be structured to cope with the most likely case of internal instability, or should it be aimed at deterring or defending against a Soviet invasion of the oil fields? This uncertainty led to a number of organizational struggles over the composition and mission of the Rapid Deployment Joint Task Force.[7]

The original chain of command called for the RDJTF to report directly to both the Commander of the Readiness Command (CINCRED) and the JCS. Struggles over service doctrine, roles, and missions added further barriers to the design of a coherent organization.[8] These organizational issues and interests were of such magnitude that they could not be resolved at the Unified Command or Joint Chiefs of Staff level. The ultimate organizational design of the RDJTF was debated by Congress and decided by the Secretary of Defense.

Thus, combatant commands designed to manage conflict in a limited contingency have had consistent and similar failures: a lack of joint and unified service participation, a struggle over roles and missions, and an inability to deploy and employ attached forces in regions of responsibility. (See table 4:3.) What was ostensibly an economical and efficient organizational structure in peacetime turned out to be a wasteful and cumbersome command in time of crisis.

The new Central Command (USCENTCOM), established in January 1983 as a successor to the RDJTF, appears to be a step in the right direction. More will be said regarding USCENTCOM and its organizational approach to structuring a limited contingency force in the concluding chapter.

Rapid Deployment Support

It has been said that "amateurs speak operations, professionals talk logistics." The task of the logistician is to determine the availability of the resources, how best these resources can be concentrated at the time and place they are needed, and, perhaps most important, how to transport and support the forces required. Over the last 20 years the acquisition of strategic mobility systems in adequate numbers to support

Table 4:3

Organizing for Rapid Deployment, 1960-1980

ORGANIZATION	PRIMARY MISSION	SCENARIO SPECIFIC	SERVICE SPECIFIC	ACTIVE GROUND FORCES AVAILABLE
STRIKE, 1963	To execute contingency missions and provide a strategic reserve of combat forces	GLOBAL	Army-Air Force	6⅓ Divisions including 82nd Airborne Division and 101st Airborne Division (Air Assault)
STRIKE, 1970	Movement coordination of all AF and Army units to SEA	GLOBAL w/focus on MEAFSA	Army-Air Force	4⅓ Divisions including ⅔ of 82nd Airborne
REDCOM, 1975	Joint training of assigned forces, reinforcement of overseas commands	NONE	Army-Air Force	5⅔ Divisions including 82nd and 101st
RDJTF, 1980	To be prepared to deploy and employ designated forces in response to contingencies threatening U.S. vital interests	FOCUS ON SOUTHWEST ASIA	All four Services	5 Divisions including 82nd and 101st + 1⅓ Marine Divisions

SOURCE: Author's estimates

rapid deployment forces has proved unsatisfactory. Shortfalls in the design, acquisition, and employment of these transport systems have in turn plagued the efforts of "half war" forces and organizations. Defense Secretary McNamara pointed out the inherent uncertainties in constructing strategic mobility systems designed to support rapid deployment: there were a variety of contingencies requiring a mix of delivery systems; the identity, and therefore the strength, of the adversary was unknown; and a range of conventional capability was required. Because different contingencies would require differing forces, a mix of strategic mobility systems was a must.[9]

The perception of the threat that engendered this rapid deployment transportation force requirement was identical to that which prompted the creation of Strike Command. The world was seen as an East-West battlefield; the prize was the allegiance of the Third World. As a consequence of such thinking it was necessary to have the capability to oppose aggression, prop up governments favorable to the United States, or demonstrate US interests across a wide spectrum of conflict and contingencies. The rapid deployment strategy was developed to meet these interests without stationing large forces overseas, thereby providing force flexibility and, ostensibly, reducing costs.

The backbone of such a strategy would be adequate strategic mobility. Theoretically, the choice lies among the concept of rapid deployment of a central reserve of ground forces, the basing of forces overseas, or some mix of the two, such as prepositioning of forces or supplies. There are advantages and disadvantages to each of these broad approaches.

The term "forward strategy" has been used frequently to denote the policy of stationing a large force close to or within the area of US security commitment, rather than holding the force within the CONUS as a strategic reserve. One of the strongest arguments in favor of forward deployment is the demonstrated resolve to defend an area. Perhaps the classic case is Korea, where the absence of US troops, coupled with the failure of the Secretary of State to include the area within the stated US defense perimeter, may have led the North Koreans to assume that this country would not come to the aid of the South. Thirty-eight years after that initial attack, despite repeated efforts by several presidents to withdraw ground forces, a strategy of rapid deployment has not been substituted for a US troop presence in Korea.

If stationing forces abroad clearly implies a commitment to the military defense of that state or region, it is unclear just how big a force makes that commitment appear credible. The "tripwire" theory suggests that only a small presence is required to trigger such a response.[10] The size of the US troop commitment to Europe has often been criticized on this basis. However, "tripwire" theories fail to consider a dynamic analysis of the opposing forces—that is, a determination of the warfighting effectiveness of the deployed force and its impact on the outcome of the conflict. In addition, such tripwire strategies tend to ignore the possibilities of a strong conventional defense and imply a rapid escalation of the conflict to the level of a "tactical" nuclear exchange.

There are several advantages to stationing forces abroad in areas deemed worthy of such military presence: added credibility and, therefore, enhanced deterrence; a capability to deal with surprise attacks; and an in-place infrastructure to cope with a major assault and to support reinforcements. The principal disadvantage is cost—certainly economic, possibly political.

Although the costs accrued in the adoption of a forward strategy appear acceptable in support of forces allocated to a major contingency, none of the options seem desirable for a limited contingency force. The stationing of forces overseas loses the flexibility necessary for a versatile force. Prepositioning on land, attractive economically, limits strategic mobility. Sea-based prepositioning is limited in application, rapid breakout, and movement. An effective rapid deployment strategy for a lesser contingency seems to require the development of strategic mobility systems.

If forces stationed abroad were judged too costly and too inflexible and if foreign policy commitments were not to be reduced, then the idea of creating a central strategic reserve of forces to meet these commitments has an enormous appeal.

Under this concept, a single force, able to deploy rapidly to any trouble spot, could meet US commitments without the costs of overseas basing.

The perceived need to make this rapid deployment capability a global one complicated force and support planning for the limited contingency force. The question became not which *single* rapid deployment strategy to select, but which *mix* of strategies appeared most appropriate given a particular scenario or contingency. Because of numerous logistical alternatives, as well as conflicting political and military requirements, this mix would not be a constant.

An added problem in determining a proper rapid deployment strategy was that, although the questions being asked in the early 1960s about support for a limited contingency force were clearly interrelated, the functional areas were organizationally compartmentalized. As Enthoven and Smith explained:

> In 1961, each of these elements—the airlift, the sealift, the bases, the prepositioned equipment, the planned deployments and the readiness—was the responsibility of a different group of people in the Defense Department. The elements were seen as separate and unrelated entities.[11]

In spite of the acceptance of the task of supporting a limited contingency force *and* the imposition of greater centralization of the systems acquisition process imposed by the new Secretary of Defense, it was not easy to achieve agreement within the Department on "how many forces we wanted to move, where we wanted to move them, and how fast."[12] Force planners faced a dilemma: "Do we want to get there quickly and in large numbers and pay the extra cost, or do we want to take our time, save money and accept greater risks?"[13]

Resolving this dilemma required, in part, determining the most efficient way to construct a rapid deployment strategy. Early systematic studies at the RAND corporation examined land-based prepositioning as a complement to airlift in the rapid deployment of forces.[14] In these studies and others, the principal tradeoffs were between the costs of prepositioning and the costs of acquiring and operating military air or seaborne transport systems. Also important, but usually unstated in these cost calculations, was the decrease in flexibility as the emphasis shifted from air and sealift toward a greater reliance on prepositioning.

A major study in 1964 by the Special Studies Group of the Joint Chiefs of Staff considered the appropriate mix of rapid response forces. The study "Rapid Deployment of Forces for Limited War," examined alternative Rapid Deployment strategies to counter possible enemy assaults in Europe, Korea, and Southeast Asia. As Enthoven and Smith surmised, the conclusions of the JCS study reflected US experiences in World War II and Korea. Rapid US reinforcement of allied armies would measurably improve the capability of indigenous forces to resist. Rapid deployment would also permit counterattacks, thereby preventing the enemy from consolidating his position. If the United States could reinforce rapidly and thus help halt the enemy advance before a significant amount of territory was captured, the war could be ended quickly at a much lower total cost.

These studies also enabled a quantitative cost criterion to be applied to force planning. In comparing the costs to fight a major conventional war with those resulting from a limited rapid deployment (assuming the larger war was deterred), planners estimated that more than $10 billion could be saved. From an economic as well as a strategic perspective the studies advocated the acquisiton of a significant lift capability for the rapid deployment of conventional forces.[15]

The JCS and the Systems Analysis office in OSD jointly constructed a mathematical model of this rapid deployment strategy. Through an iterative process the analysts systematically examined rapid deployment strategies comparing cost with capability. The recommendations of this analysis called for a mix of sealift and airlift. The analyses were, at first, based on a few relatively unsophisticated linear equations designed to determine least-cost solutions to deliver a certain tonnage of supplies within a given time. Later, over 400 equations were used to arrive at the first estimates of a strategic mobility force to support rapid deployment.[16]

Although the rational models applied to force planning for strategic mobility increased in complexity throughout the years, the conceptual problems were probably not as difficult as those in conventional force exchange models nor were they subject to the immense uncertainties of a strategic nuclear simulation. Measuring transportation costs and time is significantly easier than quantifying warfighting. However, the relatively simple problem of calculating the number of air and sealift systems required to deliver X tons of cargo in N days to Y theater Z miles distant can become considerably more complex when factors of attrition and the vulnerability of the mobility systems are introduced.

The attention to sealift as a method of rapid deployment was a new initiative in the late 1960s. Bernard Brodie as early as 1958 had remarked on the necessity of "guaranteeing the quick transportation of large bodies of men with their heavy equipment and supplies" through sealift.[17] But, just as airlift was accorded a secondary mission in the Air Force, so was sealift assigned a low priority in Navy programs and budgets. The early concentration of the Kennedy administration on airlift for rapid deployment, along with the establishment of Strike Command without a Navy component, offered little incentive to the Navy to pursue a separate mobility program for US Army ground forces that might conflict with more traditional Navy roles, missions, and budgets.[18]

However, the original Navy concept of high speed sealift to transport troops to their prepositioned equipment was expanded in a Navy study completed in 1964. "Logistic Support of Land Forces," or LOGLAND, demonstrated that seaborne transport could be vital in the deploying of ground forces. The increasing amount of heavy, non-air-transportable equipment being assigned to Army divisions made sealift appear especially attractive. A series of studies included in this report showed that a Fast Deployment Logistics Ship (FDL) could store, transport, and put ashore equipment in support of the Army and would enhance rapid deployment.

LOGLAND also developed the concept of operating ships loaded with this equipment in forward objective areas or in regions where a lesser contingency appeared likely. Additional ships in some degree of readiness would remain partially loaded in continental ports. When the order came to deploy, the FDLs could arrive at the designated area within only a few days, contributing the prestocked materiel to the troops arriving by aircraft.

This Navy recommendation for a seaborne system came at a time when the JCS Special Study Group was studying rapid deployment strategies and the services were seeking ways to implement such strategies. The FDL proposal appeared to round out a list of available alternatives, adding a sealift element to the previous mix of airlift and prepositioning. This proposal was enthusiastically welcomed by an Army that had long complained of the inadequacy of both air and sealift, as well as the lack of motivation on the part of the Air Force and Navy to procure sufficient numbers of those systems.

According to this position agreed upon by the administration, and the Army, and the Navy, the FDL would:

- Satisfy US requirements for specialized military sealift for rapid force deployment or reinforcement,
- Help deter aggression by increasing the US capability to deploy forces rapidly,
- Increase force effectiveness at less cost than a system involving only airlift and land prepositioning,
- Modernize US prepositioning afloat,
- Reduce requirements for additional prepositioning and forward deployment, and
- Possess desirable ship design features, such as high speed, ample storage capacity, and roll-on/roll-off capability.[19]

In contrast, the development of airlift doctrine in the post-World War II Air Force proceeded along four lines. The first was to use airlift to move personnel, equipment, and supplies. The second was a specialized logistic function to airlift high-value items, thereby reducing supply pipeline times, inventory, and costs, and the third was the tactical airborne operation. The strategic movement of ground forces was fourth. While the Air Force saw the first two airlift functions as most important in support of its strategic bombardment mission, the Army understandably was more interested in missions three and four, tactical airborne operation and the strategic movement of ground forces. This difference in viewpoint and mission set the stage for early organizational conflicts over airlift priorities.

The partial resolution of this conflict began with the work of the Subcommittee on National Military Airlift of the House Armed Services Committee in 1960, chaired by Congressman L. Mendel Rivers of South Carolina. The Rivers panel was charged to look into all aspects of national airlift including the Military Air Transport Service (MATS), the Civil Reserve Air Fleet (CRAF), and the effectiveness of the air transport system in terms of its contribution to national defense.[20] To deal with the legitimacy of the Army laments about the adequacy of airlift, the Committee studied JCS war plans—other committees had been reluctant to do so—to draw the proper relationships between policy commitments and the force planning to support those commitments.

The Rivers Subcommittee ended its hearings with the conclusion that existing airlift was inadequate to meet wartime needs. To overcome these limitations, the Committee made numerous recommendations, including the immediate development of a military jet cargo aircraft, the modernization of the current MATS fleet, and the upgrading of CRAF capability.[21] Thus the Air Force was given both the direction and the authorization to build an airlift force capable not only of supporting the Strategic Air Command, but also of projecting US ground forces overseas.

In response to Congressional and Army prodding and to attain increased mobility for the general purpose forces under the new strategy of flexible response, Secretary McNamara called in 1961 for a 100 percent increase in airlift capability by the end of 1964. This program requested a $172 million increase in FY 1962 for airlift, delayed the planned elimination of some C-118 and C-124 squadrons, and increased the acquisition rate for the C-130 and C-135.[22] Development of the jet transport, the C-141, was also expedited.

Again in response to the Rivers Committee, the Air Force in 1961 proposed a large jet aircraft, ''designed to accept all large general purpose forces equipment which could operate in a combat environment and substantially increase our total rapid mobility capability."[23] This aircraft, which was to become the C-5A, evolved from a specific Army requirement and derived from previous operational airlift experience. Designed to carry all of the equipment too large for transport by the C-141, the C-5A promised to end the normal practice of "tailoring" combat forces for air movement with a corresponding loss of firepower. This additional capacity, of course, implied that the C-5A would have great utility in reinforcing US troops deployed to a major contingency as well as facilitating the movement of heavy forces to a lesser case.

In 1963 an OSD study further helped convince the Air Force that the inability to deploy ground forces quickly was the chief weakness in the nation's capability to wage limited war, especially if more than one contingency were to occur simultaneously.[24] Plans were soon underway to exercise the airlift of an entire division to both Europe and the Pacific—albeit not simultaneously. However, General Adams, the STRIKE Commander, testified before the House Armed Services Committee on the need to plan and prepare for the airlift of troops to both contingencies at the same time. According to Adams, this inability to support simultaneous contingencies meant that he could not accomplish his mission and that the United States could not meet its commitments.

In September 1964 a joint Army-Air Force study called AIRTRANS-70 compared two alternative airlift forces: a 20-squadron C-141 force and a mix of C-141 and C-5A aircraft. This study found in favor of the mixed force based on its outsized capability, operating cost reduction, and potential to meet rapid deployment strategies. The study concluded that no other system or combination of rapid deployment methods—prepositioning or sealift—would provide an effective alternative to the airlift capabilities of the C-5/C-141 mix.

Consequently, after testing a wide range of various combinations of airlift, sealift, and prepositioning, the most cost-effective force to meet these requirements was determined to be 6 C-5A squadrons, 14 C-141 squadrons, 30 FDLs, equipment prepositioned in both Europe and Asia, plus a Civil Reserve Air Fleet, and 460 commercial cargo ships. This force mix had been determined analytically to meet the demands of the rapid deployment strategy to support the major and lesser contingencies that might occur.[25]

As attractive as this concept was on paper, the lift and the logistical systems to support such a strategy were never acquired. The major problem was not one of resource availability but of resource allocation. The maintenance of a ready, in-being force and the procurement of a sophisticated mix of mobility systems entailed a continuing peacetime "pre-aggression" cost that proved difficult to defend on the grounds that certain contingencies might arise.

In addition, questions were raised regarding the ultimate use of the military air and sea transport fleets in peacetime. Such capabilities appeared to infringe on the commercial territory and economic well-being of private carriers. A combination of "anti-intervention" arguments during Vietnam used against the planned-for Fast Deployment Logistics Ship and "cost overrun" statistics levied against the C-5A transport aircraft served to diminish the procurement of the optimum transport mix. Moreover, the gradual escalation of the war in Southeast Asia and the lengthy but secure lines of communication across the Pacific seemed to reduce the premium that had been placed on rapid deployment to a limited contingency.

But if rapid deployment support systems were seen as unnecessary during the Vietnam war, they were seen as superfluous after its termination. The lowered perception of the threat and the limits imposed on the defense budget in accordance with the Nixon Doctrine ensured that forces dedicated to the support of a limited contingency would lose priority. Without the requirement to deploy troops to the Third World, there was no need for a capability to do so. Thus, budgetary support for the rapid deployment methods that had survived the war would have to be based on their contribution to the support of a major contingency.

Therefore, the only strategic mobility systems that survived the transition from the McNamara formulation to the Rapid

Deployment Force of the 1980s were the airlift and prepositioning methods dedicated to the support of a NATO contingency. However, these systems remained severely limited, both in sea and airlift. When the FDL failed to win congressional funding, the C-5A became the workhorse of strategic mobility. But that aircraft had been limited, not only in total numbers acquired, but also in its specifications to meet requirements for operation in an austere or remote environment. Similarly, methods of using the Civil Reserve Air Fleet (CRAF), bare-base techniques, or land prepositioning all possessed inherent drawbacks when applied to a limited contingency. In a time of reduced spending and with a focus on Europe, these shortfalls were not regarded as particularly serious, especially given the fungibility of the systems that were expected to perform adequately, although perhaps not simultaneously with a major reinforcement effort.

Strategic mobility programs during the late 1970s were marked by fiscal austerity and a willingness to fund only those programs clearly in support of a major military effort. A RAND study completed in March 1977, "Strategic-Mobility Alternatives for the 1980s," examined only the systems required to reinforce US troops committed to NATO. In the process of projecting the airlift mix of C-5As, C-141s, and the CRAF required to meet US needs in the coming decade, the report concluded that meeting the requirements of NATO reinforcement would provide a capability adequate to serve most other conceivable needs.

A 1978 JCS study of the resources available for strategic mobility, "Strategic Mobility Requirements and Programs—1982," examined the capabilities of the existing systems and suggested that requirements for simultaneous deployments to contingencies in separate theaters could not be met. However, a Government Accounting Office evaluation of this study, while acknowledging that the JCS document represented a comprehensive examination of strategic mobility requirements and programs, questioned the budgetary justification for increased strategic mobility expenditures. Throughout the late 1970s, fiscal austerity limited funding to planning for major contingencies.

Yet the JCS study was just one example of the growing concern with the adequacy of strategic lift. A government-wide war-mobilization command-post exercise codenamed "Nifty Nugget" also generated considerable doubt. Though not an actual mobilization, Nifty Nugget was a comprehensive simulation of national mobilization conducted in October 1978. The scenario for the exercise called for a 30-day period of preparation and mobilization for an all-out conventional war against the Warsaw Pact in Europe. Although NATO-oriented, the lessons learned from this exercise had implications for improvements in mobility support systems that extended beyond a US capability to meet only the major contingency.

The chronology, events, and results of Nifty Nugget, although both interesting and revealing, are best left to other studies.[26] Most important here was the stimulus provided by the exercise to several continuing efforts to improve United States lift capabilities and the heightened perception of the importance of these systems in conducting any conventional war. Although it is not possible to draw a direct-cause and-effect relationship between the evaluation of Nifty Nugget and the support or proposal of certain mobility programs, it seems clear that a good deal of the emphasis placed on strategic mobility within the Carter administration stemmed from an analysis of the lessons of this exercise. For example:

● A new agency, the Joint Deployment Agency (JDA) was established to revise and amend existing JCS mobility plans. This agency forecasts mobility requirements for possible contingencies and makes realistic estimates of scenario air and sealift needs. The JDA, according to its mission statement, is the centralized coordinator for land, sea, and airlift in the planning and movement of forces, equipment, and supplies to support contingency operations.

● An interagency group was formed within the National Security Council to coordinate requests and correct flaws in the existing mobility plans of civilian government agencies while, at the Department of Defense, mobilization and deployment study groups processed budgetary changes and proposals to existing mobility programs.

● Prepositioning measures received greater emphasis. The program in Europe was extended, raising to six the number of division sets to be prepositioned for NATO reinforcement. Prepositioning was also seen as necessary in preparation for a lesser contingency, although land-based prepositioning appeared inappropriate. POMCUS (the prepositioning of materiel in Europe), unlike other strategic mobility initiatives that could enhance non-NATO deployments, was judged counterproductive to a capability to deploy to a lesser contingency. The requirement to preposition additional equipment in Europe was seen as likely to exacerbate existing shortages, and thereby reduce the readiness of forces that might be deployed to a non-NATO contingency.

● A series of sealift improvements were encouraged that included increasing the size of the Ready Reserve Force from 14 to 34 ships and acquiring a fleet of modern cargo ships for rapid deployment by sea.

● Airlift improvements included the C-141B stretch/refueling modification, C-5A and C-141 utilization rate increases, and CRAF enhancement. However, these measures continued to be characterized by a lack of urgency and by a NATO orientation. The CRAF program in 1979 shifted emphasis from the modification of existing aircraft that increased military cargo capability to the incorporation of these outsize-cargo modifications during aircraft production. While cost-effective, this alteration in the approach to CRAF enhancement considerably delayed a rapid deployment capability that, for the lesser contingency particularly, was already seriously constrained. The planned procurement of the Advanced Tanker Cargo Aircraft (ATCA, the KC-10) was one airlift program specifically offered to improve US rapid deployment to a limited contingency.

Later, in announcing the formation of the Rapid Deployment Force in 1979, Defense Secretary Brown stated that the proposed mobility systems for the limited contingency force owed a good deal to the Nifty Nugget exercise. That simulated mobilization revealed that air and sealift for a NATO war would rob those systems from forces designated for a lesser contingency. Thus Nifty Nugget had demonstrated—on paper—that existing US strategic mobility capability could not support the "1½ war" concept.

In this way, US strategic mobility inadequacies were well documented when the deteriorating situation in Iran, the Soviet presence in Cuba, and the Soviet invasion of Afghanistan led to efforts to strengthen the US ability to move forces rapidly to potential trouble spots. To provide for prepositioning, the Defense Department launched a program to provide a number of maritime prepositioning ships to increase deployment flexibility and avoid the problems of large permanent United States bases overseas. While these ships were being procured, constructed, or modified, an immediate prepositioning option provided a fleet of commercial-type vessels for interim storage and offload capability in the area. The plan called for the loaded ships to be prepositioned within a few days' transit time of the Persian Gulf.

Major improvements were programmed in air and sealift capabilities. Procurement of the KC-10 tanker, begun several years before, would be accelerated. The United States was also beginning a long-term program to buy a new "CX" transport aircraft for the long distance deployment of outsize cargo. To bolster sealift, the nation was acquiring high-speed civilian ships for rapid deployment by sea, although this improvement appeared as a supplement to airlift.

A third element of the new lift strategy was access to facilities in the Persian Gulf/Indian Ocean regions, including a continued buildup of US facilities on Diego Garcia. Such access would improve the ability to sustain naval and air deployments and enable the United States to aid states in the area without asking them to host or support permanent US outposts.

The mobility force programs of the late 1970s had been derived from analyses similar to the JCS Special Study Group work that had resulted in the rapid deployment mix of forces designed in the mid-1960s. Moreover, the resulting mix bore an uncanny resemblance to the rapid deployment posture that had emerged from similar studies in the 1960s (see table 4:4). Although not generally collected together in a coherent fashion, the 1980 balanced mix of airlift, sealift, and equipment prepositioning consisted of:[27]

- 70 C-5As and 234 C-141 A/B (Active)
- 50-200 CX aircraft
- CRAF (231 passenger and 111 cargo)
- 7 Near term prepositioning ships (NTPS)
- 15 prepositioning ships (MPS)
- 8 SL-7 fast sealift ships (FSS)
- 170 ships from the Sealift Readiness Program
- 43 ships from the Ready Reserve Force by 1986

Table 4:4
Mobility Systems for Rapid Deployment

January 1968	Strategic Concept	January 1981
"2½ Wars"	**Strategic Concept**	"1½ Wars"
STRIKE Command	**Organization**	RDJTF
Reinforce allies in Europe	**Purpose**	To be able to support simultaneously deployments to Europe and to other potential trouble spots
Rapidly deploy general purpose forces to counter a major conventional attack in Asia		Meet inter-theater and intra-theater requirements of a dual contingency
Meet a minor contingency elsewhere		Persian Gulf focus

Force Mix

Proposed	Acquired/Remaining	Proposed
Airlift:		
6 C-5A Squadrons		—
(120 a/c)	81/70	
14 C-141 Squadrons	224/234	—
(224 a/c)		
CRAF (465 B707/DC8)	(231 Passenger, 111 Cargo) wide-body	50-200 CX
Prepositioning:		
Europe (POMCUS)	2 Sets/4 Sets	6 Sets by mid-1983
Asia	3 FFD/0	—
Middle East	0/0	7 NTPS in Persian Gulf 15 T-AKR MPS by 1987
Sealift:		
30 FDL	0/0	8 SL-7
460 General Cargo ships	460/385	—

SOURCE: Derived from DOD Annual Reports for FY 1980, 1981, and 1982. For further elaboration of the relationship between rapid deployment concepts, organizations and purpose, see Robert Haffa Jr., *The Half War* (Boulder: Westview, 1984), Chapter 4.

5. PLANNING US FORCES

Although incomplete and imperfect, a rational basis for the planning of US military forces exists. An analytical foundation has been laid to size and structure the United States strategic, general purpose, and rapidly deployable forces. Over the last 25 years, despite some alteration in strategic concepts, these forces have remained relatively constant in size, if not in capability. In the 1980s the United States embarked on a major effort to shore up these forces which had been designed and procured more than two decades ago. The question to be addressed here is, "Are the rational procedures that helped guide force planning in the past operative and influential today?" The answer is important, for the decisions of force planners now will affect the ability of US forces to deter and defend in the next century.

Recent emphasis on American defense policy has generated an enormous amount of literature about US military strategy, tactics, and weapons acquisition practices. The concentration here will remain on the requirements for rational force planning. In a time of relatively permissive public support for robust defense budgets, force-sizing exercises appear less compelling. In a time of reduced defense spending, a basis for rational and prudent choice is required. This chapter examines the contemporary thrust of force planning in the three main categories just considered: strategic, conventional, and limited contingency operations. In each case the conclusion offers recommendations about how the planning process can be strengthened to ensure the future effectiveness of US military forces.

Strategic Nuclear Forces

The policy statements, force development plans, and weapons employment decisions made in the 1960s held through most of the 1970s. Midway through this period the United States made three important choices ensuring that its strategic force posture would continue to resemble closely the triad of forces designed by McNamara.[1] The first decision was made against a comprehensive anti-ballistic missile (ABM) system. An agreement with the USSR in 1972 to limit the deployment of such systems appeared to enshrine the declared concept of assured destruction as one of mutual interest.

Secondly, the United States elected *not* to build new strategic forces to match the Soviet buildup in intercontinental missiles, well under way in the early 1970s. Instead, from its leading position, the United States chose to accept parity or sufficiency, to limit the total number of strategic launchers on both sides through arms control agreements, and to ensure the deterrent value of its strategic forces through the modernization of each leg of the triad. Finally, and simultaneously, the decision was made not to restrict the deployment of MIRVed systems. Thus, although the size of the US strategic force appeared to remain the same, the number of warheads that could be delivered by that force—and the number of targets covered—was significantly increased.

These decisions were not made irrationally. Rather, exchange simulations acted as the guide to and rationale for force planning. As Warner Schilling has argued, considering alert rates, weapons system survivability, and weapons yield, the 1980 US second-strike force was approximately double the retaliatory force that might have been delivered in 1964.[2] The basic criteria guiding US strategic forces, that they must be capable of assured destruction and damage limitation, while appearing stable in a crisis, remained.[3] In an era of emerging strategic force parity such a policy appeared wise.

A more problematic issue attending US agreement to superpower strategic parity is how that balance has been perceived. Often the focus of the contemporary strategic debate is on the differences between the asymmetrical forces of the United States and USSR and where each ''leads'' in the ''race.'' Such a comparison, accompanied by the now familiar rising and falling red and blue bar charts, is easily absorbed by the attentive public. An acknowledgement that these static measures have not changed the deterrent balance, accompanied by a discussion of war outcomes, has been absent. Yet an emphasis on planning as opposed to perception could help restore rationality to the debate. Strategic force planners well understand that rational and fiscally realistic choices cannot be based on perception. Congress and the American public, as evidenced by the favorable votes cast on the Peacekeeper (MX) in March 1985 to strengthen the US bargaining position at the Geneva arms control talks, apparently do not.

The Peacekeeper stands as a major strategic force initiative. Proposed and accepted by both the Carter and Reagan administrations, this MIRVed ICBM reveals force planning asumptions of strategic planners over the last ten years. When the MX was first proposed in the mid-1970s as a modernization and replacement of ten-year-old ICBMs, its contributions emanated from its combination of throw-weight and accuracy, equating to a prompt hard target kill capability.

Such qualifications raised again the old strategic argument about which comes first, the strategic doctrine or the means to implement it. While the MX was faithful to employment doctrines of counterforce targeting, deployed in a fixed mode it appeared to violate declaratory guidance relying on crisis stability. Defense Secretary Brown judged that a mobile MX could unite declaratory and employment policies. And, as land-based ICBMs became increasingly vulnerable, both sides would have to face the considerable costs of making them survivable. But the deployment of Peacekeeper in fixed silos is troublesome. From a force planning perspective, such a deployment can be seen as a victory of perceived "equality" of strategic forces over crisis stability and rational planning. Where the original multiple protective shelter (MPS) scheme for basing the MX posed a significant targeting problem for the Soviets, placing the ten-warhead Peacekeeper in a fixed Minuteman silo provides a more tempting target for a Soviet first strike.

While it cannot be shown conclusively that a mobile MX deployment will greatly enhance deterrence—approximately 50 percent of US ICBM capability will remain in fixed silos, while the enormous retaliatory potential of the SLBM and bomber forces are unaltered—the rejection of mobile basing appeared to lessen the importance of crisis stability as a criterion for the planning of strategic forces. If the decision to base Peacekeeper in Minuteman silos holds, and mobile or rail-garrison basing is eschewed, it may be said that the political concept of perceived equality has triumphed over analytical measures of deterrent value.[4]

But if a *perception* of what is required for strategic parity becomes a guide to the acquisition of US strategic forces, then the analytical basis of force planning is lost. Although they may have preceded, rather than followed, the declared countervailing targeting policy of Presidential Directive 59, the strategic forces planned in the late 1970s appeared to match employment and declared doctrine while maintaining crisis stability. These forces were composed of the mobile MX, the Trident SLBM, air-launched cruise missiles attached to an aging but adequate B-52 bomber force, and a follow-on advanced technology bomber (B-2).

These substantial force improvements have been supplemented to a considerable degree by the architects of strategic planning in the 1980s. With the Peacekeeper planned to be deployed in Minuteman silos, the Scowcroft Commission called for the development of a new, smaller, mobile ICBM. Besides the follow-on ''Stealth'' bomber, the B-1 was resuscitated. In addition to the B-52 ALCM program, sea-launched cruise missiles supplement the ongoing Trident program and an advanced, ''Stealth'' air-launched cruise missile was proposed. Yet as extensive as these programs are, the perception of continued growth in Soviet nuclear arsenals and capabilities resulted in a call for the research, development, and deployment of strategic defensive systems.

The purpose here is not to question the worth—nor the costs—of these new strategic offensive and defensive programs; it is to inquire into the basis for such planning. Are the goals that were claimed by the force planning process of the 1960s and 70s—first, assured destruction and damage limitation, then strategic parity and crisis stability—now being replaced? Should they be? Are the aims of strategic defense and strategic superiority—deliberated on and put aside by the force planners in the 60s and 70s—now being reclaimed by the planners of the

1980s? Are those goals based on perception rather than analysis? How would such forces be limited?

Answering these questions requires a return to the basic policies of strategic declaration, force development, and weapons employment. Decision levels must be more closely linked if a rational force posture is to be planned and programmed. While it can be argued that the forces were planned in the 1960s on a rational (assured-destruction) basis, a considerable gap between declared and employment policy has continually characterized the planning process.

An opportunity to reconnect the three levels of strategic force planning was lost in the 1970s. There are a number of reasons for this failure.[5] Dollar constraints in the post-Vietnam defense budget played a role. Arms control agreements were seen as being of much greater value in limiting strategic forces than has been the case. The attention given to political perceptions of the balance rather than to analyses of outcomes has done more harm than good. The influence of these factors on current strategic forces has been considerable. The general tendency was to replace the strategic force on a one-for-one basis rather than to adopt a more comprehensive approach.[6]

In addition to these externalities, it has also proved difficult to draw together these decision levels in an organizational sense. Declaratory policy, enunciated by the President and the Secretary of Defense, must retain the flexibility that high level policymakers demand. And while US strategic nuclear policy at this level can be described as both consistent and evolutionary since the 1960s, presidential elections periodically interrupt the process with the review and renaming of declared doctrine. In the 80s, we have returned, once again, to a declared policy of flexible response.

Force development policy has often been separated from declaratory policy. History suggests that, regardless of the strategy declared, the armed services are apt to argue for more forces to support it. Thus under the relatively finite strategy of massive retaliation, President Eisenhower received a report that the Strategic Air Command was "manipulating calculations on the probability of target damage so as to provide a powerful argument for massive increases in SAC forces.''[7] One of the reasons McNamara allegedly backed away from declared counterforce policy was the force requirement it generated. In 1979 the Commander-in-Chief of SAC informed Secretary Brown that US strategic forces could not implement the strategy embodied in PD-59 without significant increases. The services and CINCs stepped forward smartly with their ''wish lists'' in the early 80s. Often, the ultimate arbiters of these policy debates proved to be program analysts. But decisions on force structure based on arbitrary budget goals do little to help relate forces to strategy and are hardly a rational way to design or acquire forces.

If politics dominates the first level and economics the second, it is not until we reach the third level of employment planning that the military comes into its own. That the military should rule here seems proper, as the decisions on targets, tactics, and campaign strategies require the expertise of the professional. There is also support for the position that employment policy should be the pre-eminent element of nuclear policy. Rational policymaking implies that decisions on how US forces would be used should precede decisions on force acquisition and deployment.[8] But since the time Secretary McNamara sketched a baseline strategic force based on a redundant delivery capability rather than on employment doctrine, this has not been the case.

Moreover, employment planning is probably the least understood and the most autonomous of the three policy levels. The blame must be shared. Until Richard Nixon received a SIOP briefing in 1969, leading him to provide better presidential guidance on strategic force employment, planners had been required to infer employment policy from administration declaratory policy. But force capability far exceeded the requirements of a countervalue second strike. Defense Secretary Schlesinger ultimately provided explicit guidance in the form of requirements for ''Limited Nuclear Options,'' and Jimmy Carter was probably the first President to consider seriously his role as Commander-in-Chief of a nuclear warfighting force.[9] On the other hand, nuclear warfighting documents, as employment plans, have been closely held by the strategic planners.

A rational basis for the planning of US strategic forces therefore calls for greater cohesion in US declaratory, development, and employment planning. President Carter's PD-59 suggested that US declaratory policy was then approaching, if not directing, the targeting plans that have included limited nuclear options and attempted escalation control for more than a decade. As Desmond Ball has pointed out, better connectivity between declaratory and employment policy demands improvement in the existing machinery. Employment planners must pay increasing attention to policy pronouncements such as the Nuclear Weapons Employment Plan, while decision makers must appreciate the uncertainties that dominate the domain of the strategic targeters.[10]

A better fit between public policy and targeting doctrine can reveal the difficulties of planning strategic forces, but it cannot determine how much force is enough. Force programs in the Carter and Reagan administrations reflected differing judgments and perceptions about how the United States can best plan its forces to assure the credibility of deterrence while including a warfighting capability. Shrinking defense budgets will demand that those planning methods be revisited.

A number of approaches can help rationalize the strategic force planning process. The *first* step—reconnecting policy formulation and employment planning—is already underway and should be continued. The *second* step should be an examination of the systems themselves. Strategic weapons systems should be planned and procured according to their deterrent value in contributing to the overall US strategic posture, not with the principal concern being how such an acquisition will appear to our allies or our adversaries. History provides good evidence of the limited worth of nuclear weapons, while analytical and hypothetical force exchanges provide us with an empirical base on which to make a reasoned judgment on the size and type of these forces.

A *third* principle is to emphasize planning rather than budgeting. Just as the planning of strategic forces must be divorced from perceptions, so also must it be separated from the programming and budgetary cycles of the process that will tend to dominate it if left unchecked. The defense budget should be an output of the process, not an input to important strategic decisions. Just as throwing money at the problem does not guarantee that the force generated will be desirable or suitable, the paring of the defense budget by some fixed or compromised percent in order to reach a fiscal rather than a security goal undercuts the rationally planned force. This is not to imply that the planning of strategic forces should be immune from program/budget cycles, but rather to insist that choices between a minimum risk force and existing capabilities are made explicit in terms of reduced expenditures and increased risk.

A *fourth* principle to be emphasized is the need for long-range planning. When one considers that both the MX and the B-1 were proposed in the early 1970s, it becomes apparent that major force acquisition efforts *are* long-range projects. The adaptability of the Minuteman, Polaris, and B-52 forces over two decades should be kept in mind as the United States plans its forces in the 1980s. These new forces, too, will likely be required to last longer and to be deployed later than originally planned.

These four themes—coherent policy relationships, a reliance on empirical rather than data-free analysis, emphasis on planning as opposed to budgeting, and a need to look to the long term—can serve as guideposts in the planning of a strategic force that genuinely contributes to greater US security.[11]

General Purpose Forces

Chapter 3 of this study reflected a rational relationship between US commitments abroad and the planning of general purpose forces. In comparison, this planning process is similar to the three-tiered decision levels that act in the planning of strategic nuclear forces. In the conventional case, declaratory policy is composed of the defense commitments the US has made to its allies and the strategic concept that defines these interests for planning purposes. Force development planning results from an assessment of the forces needed to support those agreements; employment planning designs war plans to execute the strategy.

The actors involved in this planning process also parallel the three nuclear policy levels. Declaratory policy continues to lie at the highest levels of the government, although here the legal and moral continuity of US commitments allows the President and his advisers less opportunity for reappraisals of US interests. The military again rules at the third level where wartime tactics are planned. While the middle ground of force planning has often been occupied by the systems analysts in OSD, their influence has not proved as great here as in the nuclear realm. Conventional force exchanges are more difficult to quantify than atomic ones, and the menu of options is considerably more complex. Moreover, with the general purpose forces absorbing almost 80 percent of the defense budget and capturing the organizational essence of the armed services, the domestic stakes in this game are considerably higher. Little wonder that case studies in weapons system acquisition frequently turn out to be after-the-fact accounts of bureaucratic political maneuvering.

This diversity found within the conventional force planning process is not unique to the 1980s. The methodology, however, seems to have changed. The traditional way in which the United States matched conventional objectives and resources was through a series of analytical force sizing exercises such as those outlined earlier. By allowing the testing of planned conventional forces against anticipated threats in given contingencies, these quantitative methods helped planners judge the appropriate size of the forces to deter and defend. As the United States began a major general force-planning initiative in the early 1980s, some of these time-tested practices were set aside. This section inquires into the wisdom of that appproach.

At first glance it seems that contemporary conventional planners are not deviating significantly from the course set by their predecessors. Efforts continue to enhance the sustainability of ground and air forces deployed forward in Europe and Asia, both through the buildup of war readiness materiel and with the allocation of naval forces to defend the sea lanes of reinforcement. But if the practice is consistent with the past, the theory tends to diverge. In the fashioning of a new strategic concept, defense planners in the early 1980s called for the preparation of US general purpose forces to fight a "global war" against Soviet or proxy aggression on several fronts simultaneously.[12]

According to this formulation, the "mechanistic assumptions" calling for forces to be planned to fight "2½" or "1½ wars" neglected both risks and opportunities. The risks were that US forces planned for known contingencies would be unprepared to meet unforeseen ones. The opportunities were that these forces properly equipped could engage in counteroffensive operations in areas not of the adversary's choosing—the so-called horizontal escalation concept. [13]

Although some of this rhetoric has been softened in subsequent and sequential Secretary of Defense *Annual Reports to Congress*, this less structured approach to general purpose force planning has had an effect. One organizational result has been the decentralization of the planning process. In the last few years the employment planners have been able to generate force requirements for the contingencies they conceive of as likely and important. With contingency guidance from above given less emphasis, the planning profile of the OSD systems analysts was lowered and the planning responsibilities were passed to the armed service staffs.

If the military services have been granted a larger role in the planning process, it is useful to consider the effects of employment planning on force development. A rational model would consider an employment input valuable in deriving a prudent and practical force. But the force resulting from the process does not seem to have those properties. Seeking planning guidance from the bottom up acts to turn the traditional strategy-force mismatch on its head. Now the problem may not be insufficient forces to meet the strategy, but too many strategies. Without a meaningful process that makes enforceable joint judgments on force priorities and contingency planning, the services may well pursue separate and unrelated doctrines that enhance their organizational interests and support their desire for larger forces. Thus, the Navy is concerned primarily with attacking enemy naval forces, the Air Force with defeating the enemy in the air, the Army with the land battle and the occupation of enemy territory, and the Marine Corps with whatever is left— beachheads and islands. Such a simplistic description of service interests and implied interservice rivalries can be patently unfair if taken to the extreme. Yet it also can contain an element of truth, particularly if service strategy is driven more by competition for a share of the market than by some higher calling of grand strategy.[14]

A former member of the Joint Chiefs observed that interservice rivalry declines as defense budgets rise. But times of relatively unconstrained defense dollars have been measured in moments. It ultimately falls to the JCS to prepare contingency plans for unified action and make the hard choices between a planning force that would meet the threat with acceptable risk and the programmed force which must make prudent choices with limited resources. These choices the JCS has been reluctant to make, as documented by a series of former Defense Secretaries and retired Chiefs of Service.

Many reasons exist for this situation—and as many proposed solutions to the problem. A 1985 work chastised the JCS for being unable to rise above their individual service interests to provide cross-service counsel on effective force planning. The result, the study concluded, is "diluted advice wedded to the status quo, reactive rather than innovative."[15]

If the JCS remains unwilling or unable to make the hard choices and the systems analysts in DOD have abandoned the force planning arena, then conventional force planning in the 1980s has been delegated to the services. But contingency planning has lost its primacy as the methodological mainstay of force planning. What are the services using in its stead? How rational are the force planning practices in the 1980s?

Looking first to the ground forces, we might expect that the major contingency in Europe would continue to drive US Army force planning. However, the majority of force planning in the Army appears to be devoted to the creation of the "Light Division"—a unit oriented to other-than-NATO contingencies. The construction of the light division makes the traditional index of Army combat strength, the division, less meaningful. For although the number of divisions in the Army recently rose from 16 to 18, the number of personnel in that service remained capped by Congress. Thus the new light divisions are composed of about 10,000 men rather than the 18,000 or so that make up a standard division.[16]

It is too early to judge the contributions of the light division to the existing force. Acording to former Army Chief of Staff General Wickham, the light divisions were formed to meet a need for highly trained, rapidly deployable light forces.[17] But the manner in which these forces will be used is less clear. Certainly a smaller division, with less heavy and outsized equipment, will be easier to move. Where it will move to, and how, is another question. The light division is touted as a multi-purpose force capable of fighting in a range of contingencies: from low to high intensity, from mountains to urban areas, from quick raids to anti-armor tactics, from Europe to Southwest Asia. Perhaps this force can be used in all of these events, but past US attempts at creating such a flexible organization suggest otherwise, and unified commanders have expressed their doubts. It has proved more profitable to plan forces for specific contingencies and to train, equip, and exercise them accordingly. More will be said about this approach in the following section.

The major concentration of Air Force planning has been placed on the NATO-oriented strategies known variously as "Airland Battle," "Follow-on Force Attack," and "Deep Attack."[18] These plans to attack and disrupt enemy reinforcements in rear echelons place a premium on the roles of counterair and interdiction rather than on close air support. The force planning implications are for long-range, high-speed aircraft armed with precision guided munitions as opposed to close air support systems that might make up for lightened army firepower in the target-rich environment at the forward line of troops. Both missions are important. If an effort is not made at some planning level to integrate strategy and forces, the separate services will tend to plan forces for different wars, even in the same theater.

To counter this inclination, a 1984 Memorandum of Agreement between the Chiefs of the Army and the Air Force framed a series of initiatives in joint force development, including the study of certain roles and missions and cross-service participation in program design. The Memorandum was labeled historic—perhaps in recognition that little had been accomplished in the arena of Army-Air Force agreement since 1965.[19] But while that has been too long between drinks from the fountain of joint doctrine, and while the hard issues in the document have yet to be reconciled, the sheer existence of such an effort gives encouragement to those who plead that rational general purpose force planning must become a joint undertaking.

Unfortunately, talk of unified doctrine and joint initiatives often is drowned out when it comes to force planning. The Navy, long an opponent of contingency-based force planning, has taken the ball of decentralized management and attempted to run with it to the score of 15 carrier battle groups and 600 ships. Chapter 3 suggested the rationale behind such a force structure. It is necessary only to note here that there is no national policy requiring a naval presence in certain oceans or dictating the number of carriers to be deployed in peacetime. If the services are allowed to plan their own campaigns and contingencies, peacetime deployments have a way of becoming justification for larger warfighting forces.[20]

Is the right number for carrier battle groups 12, 15, or 24? The answer may depend on the credibility of rational force exchange models, the political power of the Secretary of the Navy, and the dollar amount that Congress decides it can afford. But a strategic issue larger than the budgetary one lies just beneath the bow wave of force modernization and acquisition. Former Under Secretary of Defense for Policy Robert Komer has argued that early budgetary commitment to the 600-ship Navy may result in fiscal shortfalls for other defense needs. Decentralization of force planning may not cause four or more incomplete strategies, but it certainly has occasioned two different approaches. Komer has identified these as the "Maritime Strategy"—a unilateral, sea-based policy that emphasizes naval forces and attacks—and "Coalition Defense"—an air-land battle linked with allies to deter and defend in Europe and Southwest Asia. The strategies need not be mutually exclusive. The concern is that they can become so if the rush to a naval buildup robs NATO-oriented air and ground forces and rapidly deployable forces of their share of the resources. Backing into a maritime strategy, Komer warns, may be the consequence of rejecting coherent contingency planning and allowing piecemeal planning by the services.[21]

Given the above appraisal of contemporary force planning trends, the perceived need for change may depend on the colors of one's old school tie. While the Army is lightening its divisions to deploy to out-of-area contingencies, the Air Force is seeking high-tech systems to attack the enemy supply lines in East Europe. Meanwhile the Navy is preparing to sail in harm's way against enemy coastal defenses, far from the central fray. Surely some of these capabilities are desirable; they certainly stand a chance of confusing the adversary. But the nation cannot afford a force planning process that favors individual services or irrational methods. The solution to the strategy-force mismatch, particularly under balanced-budget legislation, is not separate strategies and more forces.

Like some other remedies, the medicine prescribed to restore rationally based planning to US general purpose forces may be as difficult to administer as to swallow. One also wonders at which end to begin the treatment. Starting at the top, it is clear that a stronger link between policy commitments and employment planning is called for. Disregarding DOD management style, there is a continuing need to make difficult choices on the number of contingencies for which conventional forces should be prepared. Disregarding service parochialism, no substitute is available for the analysis of plausible campaigns in areas of vital interest to help develop capable and affordable forces.

As in the strategic case, there is a need for greater dialogue between the policymakers and the war planners. A 1980 attempt to subject JCS contingency plans to review by the Under Secretary of Defense for Policy might have educated the civilian leadership on resource shortfalls, while updating the combatant commanders, or CINCs, on declaratory policy.[22] That opportunity was lost with decentralization; recent reorganization legislation has attempted to regain it. But when the Secretary of Defense loses influence over the planning process, a fracturing and a fractioning of planning power within the Defense Department results. This fragmenting brings with it the ascendancy of service-oriented programs and budgets.

There is, of course, a legitimate role for the services and the JCS in the force planning process. That these agencies' employment plans should be influencing force planning is proper. That the influence appears to take the form of championing service agendas rather than supporting unified war plans or joint guidance is at best inefficient. Strengthening the role of the CINCs in the planning and programming process is a positive trend that must continue. Also heartening is that the JCS and unified commanders are placing greater emphasis on analytical capabilities and long-range planning.

Such incremental steps are likely to prove inadequate in reducing the overwhelming influence of the services and incapable of adding to the joint responsibility for force planning from a national perspective. One answer to this dilemma is, of course, JCS reform, and the recent passage of the Goldwater-Nichols DOD Reorganization Act offers some hope of a strengthened joint input to force planning. But rationality cannot be legislated and, although this new emphasis on joint action is welcome—as are a number of DOD initiatives such as a two-year defense budget—the results of such reforms lie in the long term.[23]

In the near term we must return to the erstwhile conventional force planners—the systems analysts residing in OSD. Their task is to make analysis more relevant to force planning. Richard Kugler has suggested that analysts place a greater effort on the achievement of military goals, construct more dynamic force balance assessments to aid cross-program evaluation, and involve themselves deeply in military operations and strategy.[24] Such an effort would help complete the loop between general purpose force policy, employment, and planning.

Certainly the force planning process would be strengthened by more effective linkage between policy and war plans, by a more unified military voice, and by better analysis and long-range planning. But the serious business of general purpose force planning requires high level decisions on appropriate forces to deter and defend in regions vital to US interests. In the end, there appears to be no substitute for a strong Secretary of Defense to restore order and rationality to the conventional force planning process.

Rapidly Deployable Forces

Recent events in the Persian Gulf, including an apparently inadvertent Iraqi attack on a US warship, the "reflagging" of Kuwaiti tankers and the accidental shoot-down of an Iranian Airbus by a US vessel in the gulf, have again focused attention on rapidly deployable forces. The planning of general purpose forces to meet a limited contingency resulting from increased hostilities or commitments in Southwest Asia or elsewhere can profit from lessons learned in the past. Chapter 4 of this study reviewed strategy, organization, and support in force planning for the "half war." From an examination of those unsuccessful experiences, the requirements for the planning of a coherent limited contingency force appear to be as follows:

Strategy. If the region is judged worthy of committing US troops in its defense, planning must be devoted to designing a force capable of defending these interests. A focus on a particular region facilitates force planning by allowing forces to be sized against a specific threat, to be trained in appropriate tactics, and to be exercised within the region. Forces planned under strategies that failed to discriminate among service interests and instead called for versatile organizations and forces were denied these advantages. Thus a strategic concept that seeks a limited contingency capability should not be interpreted as demanding a global half-war deployment capability. While the United States must face the prospect of meeting mutiple contingencies simultaneously, it must also decide which of those contingencies are most important to US interests and plan accordingly.

Organization. It follows from the strategy that a coherent limited contingency force must include multi-service forces under a unified command. Inclusion of each component in the limited contingency force does not require that all units from each service be employed in every situation. The contingency force, however, must have the capability of operating flexibly in the region and sequentially employing its own forces under centralized command and control.

Support. A limited contingency force must possess organic or dedicated air and sealift, a program of appropriate prepositioning, a means to gain access to needed facilities, and a power projection capability. If the contingency is judged worthy of defending, forces assigned to the command must be guaranteed adequate strategic mobility systems, even during a major reinforcement of another region.

If these are the half-war force planning lessons, how are they being applied in the 1980s? The strategic concept governing the planning of general purpose forces was just discussed. It appears to contain both advantages and disadvantages for the planning of rapidly deployable forces. In the first case, the United States has explicitly recognized the goal of developing a capability for deploying forces to several contingencies simultaneously. Thus, the limited contingency seems to have been rescued from its secondary status as a lesser included case. This revision in strategic thinking was based to a significant extent on a reappraisal of the threat. If the Soviets were capable of military intervention in regions outside the more commonly conceived regions of US-USSR conflict, then the United States was forced to respond. Southwest Asia has been central to this concern after events in Iran and Afghanistan.

But realizing that limited contingencies can escalate into high-intensity US-USSR conflict is only the first strategic step. Forces now need to be planned and allocated in support of those contingencies. Here the new strategic concept seems to short-change limited contingency operations. Claims about a global threat, the possibility of a prolonged conventional war, and the deterrent attractions of horizontal escalation do little to aid a coherent or rational force planning process. The concentration of general purpose force planning on Europe and the Pacific in the 1960s and 1970s led to ineffective rapid deployment forces. And although a military buildup across the board also enhances the capabilities of rapidly deployable forces, a rational plan for which forces will fight where, and against whom is also required.

Turning to the organization of rapidly deployable forces in the 1980s, it might have been expected, given the strategic concept just described, that a Rapid Deployment Force would be modeled in the image of Strike Command—a go-anywhere, do-anything force. Even though it is not clear that this concept has been completely abandoned, it does not apply to the establishment in January 1983 of Central Command. USCENTCOM is composed of all four services with a single unified commander. The creation of the Central Command required a change in the Unified Command Plan and the assignment to USCENTCOM of an area of responsibility previously shared by the US European Command and the Pacific Command, as well as some areas previously unassigned. One principal advantage of having a single command in the region is that the countries affected are asked to deal with only one organization on most security issues.

The allocation of the Southwest Asia region to a single unified command appears as a rational geopolitical choice, but there was a bureaucratic rationale as well. The Rapid Deployment Joint Task Force (RDJTF) had been plagued by conflicting and controversial command arrangements. The separate services were reluctant to surrender units earmarked for other areas of responsibility to a contingency task force. As a unified command, USCENTCOM's span of control is more definitive than that of the joint task forces, although control of its forces could be characterized as somewhat looser than that possessed by the RDJTF.[25]

Although the RDJTF commander was given day-to-day operational control over certain Army and Air Force units, the USCENTCOM Commander has forces available for planning purposes only.[26] However, USCENTCOM has access to a reservoir of forces that could be assigned depending on the nature of the contingency. Although Central Command's control over its to-be-assigned force is not substantially different from other unified commands depending on CONUS-based reinforcements, USCENTCOM is unique in that its headquarters and its component commands are not located within the area of responsibility. (See table 5:1.) Also, like other unified commanders, CINC-CENTCOM has to live with the fact that the daily operations and training allegiances of his component commands lie with their separate services rather than with him. But the presence of these components clearly enhances USCENTCOM's capability

Table 5:1

Combat Forces Earmarked for the US Central Command

Service and type	Number
Army	
Airborne division	1
Airmobile-air assault division	1
Mechanized infantry division	1
Infantry division	1
Marine Corps	
Marine amphibious force[b]	1 1/3
Air Force	
Tactical fighter wing[c]	7
Strategic bomber squadron (B-52)	2
Navy	
Carrier battle group	3
Surface action group	1
Maritime air patrol squadron[d]	5

[a]Although these forces are described as being initially available to USCENTCOM, that would be true only if no other major contingency had previously occured.

[b]A Marine amphibious force typically consists of a reinforced Marine division, a support group, and a Marine air wing (containing roughly twice as many tactical aircraft as an Air Force fighter wing as well as a helicopter unit).

[c]Typically 72 fighter and attack aircraft.

[d]Usually consists of 9 P-3 long-range antisubmarine warfare aircraft.

SOURCE: *Department of Defense Annual Report, FY 1987, p. 272.*

for mission accomplishment and encourages interservice cooperation to a greater extent than existed under the old RDJTF.

It is somewhat surprising that force planners favoring the maneuverability of a maritime, as opposed to a coalition, strategy would move in the direction of a unified command with an assigned area of responsibility. More than an organizational or strategic rationale, it was probably the perception of the threat in the region that led to the multi-service composition of the force: the scenario envisioned to guide force planning was increasingly Soviet centered. The primary mission of USCENTCOM is to deter Soviet aggression and to protect US interests in Southwest Asia.[27] However, this is not the most likely case, but only the worst. The Central Command is also engaged in planning for more likely lesser contingencies. Focused planning on a key region, undertaken by a well-staffed joint headquarters, may be the most significant contribution of CENTCOM to US force planning for a limited contingency. Most revealing of CENTCOM'S role will be its continuing control over US forces deployed to Southwest Asia to enhance US presence and protect freedom of navigation in the Persian Gulf. Keeping these forces under the operational control of CENTCOM would signify a significant departure from past practice. The more likely event of such forces remaining under their peacetime CINC/service control would suggest that little joint organization reform of contingency forces has occurred since the early days of STRIKE Command.[28]

In any case, the problem of force availability remains. Forces that could be assigned to USCENTCOM might also be assigned elsewhere and are therefore trained and exercised in other contingencies. Barring the creation of new and separate forces for each unified command—an unlikely event in the face of current resource constraints—steps to free forces for deployment to Southwest Asia will be scenario-dependent. The creation of Central Command therefore moves away from the unattainable organizational goal that characterized US planning for limited contingencies in the past—a central reserve intended to respond to any global contingency. But in assigning the same units to do multiple tasks, the issue of force versatility and concurrent availability has not been resolved.

In addition, two recent organizational reforms may hold some promise for the deployment and employment of limited contingency forces. Goldwater-Nichols legislation removed the prohibition against the establishment of a unified Transportation Command, and responded to the President's Blue Ribbon Commission on Defense Management which had urged the establishment of a "single unified command to integrate global air, land

and sea transport.''[29] The new Command will absorb the deployment planning functions of the Joint Deployment Agency, which will be phased out over a two-year period. Like the operational commands that attempted to guide limited contingency deployments before it, the JDA never had the power to carry out its responsibilities and ultimately surrendered to the service transportation commands. This new organization, along with increased attention to deployment goals, the attention of a new CINC and the attendant budgetary participation now granted to CINCs, may bring increased coherence to the deployment of US military forces to all contingencies.

The United States Special Operations Command is the second new unified command resulting from reorganization legislation, taking the place of another outdated attempt at organizing for rapid deployment, the Readiness Command. USSOCOM's active forces include the SOF units of the 23rd Air Force, the Army's 1st Special Operations Command and the JFK Special Warfare Center, and the Navy's Special Warfare Center. Established by Congress over some service opposition, the Special Operations Command was designed to guarantee the budgetary support to special operations the Congress demanded and the services granted grudgingly. But there is danger in confusing the missions of special forces with those of counterterrorism, low intensity conflict, and lesser contingencies. Special forces are structured based on the warfighting objectives of the theater commanders and are not planned to be employed primarily in very low intensity, counterterrorist kinds of operations. The new SOCOM must not be perceived as a STRIKE Command.

Finally it should be admitted that this focus on USCENTCOM and Southwest Asia should not be interpreted as the final resting place of US rapidly deployable forces. The unified command approach appears to be the best for planning the use of US general purpose forces in regions of vital interest. But rapid deployment forces—in small letters—still exist in the more traditional guises of the Army's Airborne and Air Assault Forces and the Marine Corps. Other units pledged to the Central Command could also be deployed to regions far from Southwest Asia. Thus, while USCENTCOM probably never will be directed to deploy its forces to contingencies outside its area of responsibility, the forces themselves are vulnerable to out-of-area assignments. In this regard, the Central Command can be seen as a case study for the organization and management of limited contingency forces, but not as a panacea for rapid deployment requirements.

Despite this overall favorable rating given the organizational side of planning for limited contingencies, particularly when compared with the less-capable RDJTF and JDA, the formation of organizations is far easier to accomplish than the rapid deployment of general purpose forces. A rapid deployment capability remains fundamental to a contingency force, particularly if the model of the contingency in the 1980s is Southwest Asia—where US forces most likely will not be pre-deployed. In the case of rapidly deployable forces, advances in strategic concept and organization have far outpaced the modes and monies available for strategic lift. Those planning rapid deployment forces in the 1980s have had to be content with the classic forms of projecting power overseas: sealift, airlift, pre-positioning, and access to foreign facilities.[30]

Sealift

Force planning for sealift was given focus by the Congressionally Mandated Mobility Study in 1981. On the basis of a two-war scenario, analysts concluded that one million tons of sealift capacity would meet the immediate needs of a European and Southwest Asian contingency, and would be able to sustain forces in both theaters for more than 30 days. Yet government-controlled shipping appears to be able to reach barely more than half of that goal by the early 1990s, and the prospects for commercial sealift making up the shortfall are dim.[31]

This is not to say that the issue has been avoided, The Navy has elevated strategic sealift to one of its three principal functions, now sharing top billing—if not budgetary clout—with sea control and power projection. A bureaucratic push was also provided in 1984 with the establishment of a Strategic Sealift Division in Naval Headquarters. This policy initiative has resulted in significant gains: a Ready Reserve Force of 83 ships, 13 Maritime Prepositioning Ships, and the purchase and conversion of eight SL-7 Fast Sealift Ships. But the sealift force is still regarded as ''marginally inadequate'' by those in the know, and improvements are still required to meet the demanding goals for lift if a major conflict and a limited contingency occur simultaneously.[32] Much will depend on the continued success of pre-positioning programs in Europe and Southwest Asia, the future of the US Merchant Marine, the continued interest and largesse of Congress, and the influence of the Unified Transportation Command.

Airlift

As it laid down a marker for sealift, so the Congressionally Mandated Mobility Study also established a goal for airlift—66 million ton-miles per day. And as in the case of sealift, the airlift shortfalls that remain belie the significant progress made throughout the 1980s toward that goal. For although the goal remains elusive—it will not be reached even with a full acquisition of the C-17—such aggregate measures of capability do not provide the force planner with the specifics of the contingency he needs to provide for, and do not identify the tradeoffs between intra-and inter-theater airlift, nor the demands for an airlifter capable of operating in austere low-intensity environments.

Meanwhile, procurement of C-5Bs and KC-10s, along with modifications to the C-141 and C-5A fleets have boosted airlift capacity considerably. The planned advantages of the C-17—its great maneuverability on the ground when compared with the C-5, its direct delivery capability and its accessability to smaller, shorter runways—make it the ideal airlifter for out-of-NATO contingencies. But its hefty price tag makes the C-17 a likely target for delay and program stretch-out in a defense budget seeking substantial near term savings.[33]

Prepositioning

In Europe the equation is simple—unit equipment that is not prepositioned forward will add to the above-mentioned air and sealift requirements. In Southwest Asia the problem is more complex owing to the limited facilities available for the prepositioning of US equipment. Nevertheless, some prepositioning of Army and Air Force equipment is occurring and ongoing negotiations may yield a better situation still.

In the meantime, a significant amount of prepositioned equipment continues to be maintained afloat, with twelve Near Term Prepositioning Ships still serving the Army, Navy and Air Force at Diego Garcia. The new Maritime Prepositioning Ships are there also—as well as in the Atlantic and Pacific—in support of Marine Expeditionary Brigades. The phrase in *DOD's Annual Report to Congress* that the ''MPS brigades can be rapidly deployed to any trouble spot worldwide'' suggests that the spirit of a ''go anywhere, do anything'' force once supported by the FDL still lives.[34]

Access to Foreign Facilities

Negotiations with key nations, particularly those in or near the Southwest Asia region, and arrangements to preposition materiel, conduct training exercises, and use regional facilities in the event of a crisis have continued. The emphasis here is not on constructing new bases or raising the visibility of the United States in the region, but on improving host-nation support and infrastructure. A most notable change of attitudes has been seen within the Gulf Cooperation Council as a result of US air and naval forces escorting shipping in the Gulf. Careful management of this delicate relationship could allow not only increased access to the region but also a forward headquarters ashore for CENTCOM.[35]

* * * * *

This brief review of planning for limited contingencies has served only to reemphasize lessons learned in previous decades. The strategic concept that must guide limited contingency planning remains faithful to the larger sphere of general purpose force planning. A strategic concept too narrowly drawn can invite aggression, but too grand a concept will be forced to rely on the shadow, rather than the substance, of power. For limited contingency forces, too, hard choices must be made about the importance and the affordability of specific capabilities. When the limited contingency focuses on a specific region and a particular adversary, forces can be designed for the threat, tailored to the area, and exercised appropriately. The command structure that has proved over time to be the most effective in leading multi-service forces is the unified command.

But whatever the strategy and organization, the support system is likely to determine the winner on the battlefield. Along with the need for greater and dedicated strategic mobility are also requirements for political, logistical, and military support from the states in the region. Therefore, the formulation of a multilateral strategy that weaves together programs of host-nation support, foreign military sales, and security assistance should accompany U.S. unilateral force improvements and plans.

Ultimately, of course, it will not be an abstract strategic concept, a single organization, or improved mobility systems that alter, direct, or transport US rapid deployment strategy in the 1990s, but the willingness of the American government and people to support the force structure required to meet US security interests. To earn that support, force planning for a limited contingency must remain sound and rational. Until the end of this century, at least, the planning of US general purpose forces to meet a limited contingency will play an important role in supporting commitments abroad. The strategic, organizational, and logistic experiences of the past have generally informed us how not to go about it. But steps taken in the 1980s show more promise. The challenge is to continue the effort of constructing coherent limited contingency forces that match rational deployment strategies, organization, and support.

Reemphasis on Rational Force Planning

In summing up the requirements for planning rapidly deployable forces, the above paragraph is relevant to the conclusion of this study. Ultimately, force planning will be constrained by the willingness of the Congress and the American public to support the force structure requested by the Department of Defense and the Executive branch. But the halcyon days of the early 1980s have passed and the public mood has altered perceptibly. And while the US treasury check made out to force planning is not likely to be cancelled, it is likely to be delayed in the future owing to insufficient funds and lengthy questioning of credit references. "The same factors that led Congress to reduce the defense budget for fiscal years 1986 and 1987 are all present again this year;" writes Bruce MacLaury, "it is difficult to forecast how long this cycle will last."[36]

Such a fiscal environment argues all the more for a return to the rational force planning approach outlined in this study. *Return*, however, is probably not the appropriate word. This book was not meant to imply that planners and analysts are not working on the problems suggested here. They are, armed with tools far more sophisticated than those outlined in this primer. It is a *reemphasis* on rational force planning that is needed.

To stop and listen for rustlings that may suggest a reemphasis of rational force planning methods seems prudent at this point. There are plenty of issues that demand its application. In the strategic arena the plans have been made but the modernization of the force requires several years of continuing effort. Exchange calculations demonstrate that the greatest shortfall remains in the US capability to respond promptly against hardened Soviet ICBM silos, launch control facilities, and command and control elements. Despite its checkered history, the Peacekeeper ICBM still appears to be the lowest cost, most effective approach to filling that shortfall. The application of rational methods should assist in the hard choices that will have to be made on Peacekeeper basing and among competing strategic systems constrained by arms reduction agreements.

Achieving stable deterrence at the strategic level does not relieve force planners of the need for a strong conventional deterrent at the theater level. Now, particularly as we see progress toward reducing intermediate and shorter range nuclear forces, adequate conventional forces become even more important in deterring Soviet adventurism. Considerable progress has been made; weighty decisions remain. How should ground force responsibilities be allocated between Central Europe and Southwest Asia? What is the correct fit between the Air-Land battle and maritime strategies—should we be stretching our procurement of tanks to allow an earlier buy of aircraft carriers? Rational force planning methodologies can illuminate the consequences of such choices.

In the realm of the half war, organizational efforts continue to outpace operational concepts and logistic capabilities. Targets set by the Congressionally Mandated Mobility Study—initially fiscally constrained—may have to be revisited in even less permissive budgetary climes, and tradeoffs among airlift, sealift and prepositioning must be continually reevaluated. Meanwhile, misguided proposals advocating the withdrawal of US troops from Europe to be absorbed into a new Rapid Deployment Force, or fashioning the new Special Operations Command into the latest version of STRIKE command can be enlightened by analysis and force planning history.

Rational force planning has been used to establish the baseline force and to test the adequacy of the existing force. The real question is: Can these methods be reapplied and reemphasized to plan future US military forces? Our security and solvency may depend on the answer.

ENDNOTES

1. A Primer on Force Planning

1. Paul H. Nitze, "Arms, Strategy and Policy," *Foreign Affairs* (January 1956), pp. 187-198.

2. Donald M. Snow, "Levels of Strategy and American Strategic Nuclear Policy," *Air University Review*, Vol. XXXV, No. 1 (November-December 1983), pp. 63-73.

3. The intellectual roots of the PPBS are developed by Alain Enthoven and K. Wayne Smith in *How Much is Enough?* (New York: Harper and Row, 1971). Arguments favoring and opposing the system are collected in *An Analysis and Evaluation of Public Expenditures: The PPB System* (Washington: GPO, 1969). For an inside-the-building look at the Air Force process see *A Primer, The Planning, Programming and Budgetary System* (Washington: DCS/Programs and Resources, Department of the Air Force, January 1987).

4. Lawrence J. Korb, "Planning, Programming and Budgeting System" (mimeographed document, 1983).

5. Two recent examples are Eliot A. Cohen, "Guessing Game: A Reappraisal of Systems Analysis," in Samuel P. Huntington, ed., *The Strategic Imperative* (Cambridge: Ballinger, 1982), pp. 163-192, and Stephen Rosen, "Systems Analysis and the Quest for Rational Defense," *The Public Interest* No. 76 (Summer 1984), pp. 3-17.

6. For an argument that the "McNamara Revolution" failed to achieve drastic improvements in DOD organizational control, see Arnold Kanter, *Defense Politics* (Chicago: University of Chicago Press, 1975).

7. The planning of tactical or theater nuclear forces will not be addressed here. While a rational basis exists for the planning and deployment of some of these systems—such as the 108 Pershing 2 and 464 ground-launched cruise missiles once planned for Europe—the major purpose of these systems has been political: the linking of

NATO's conventional defense with US strategic nuclear forces. The recently signed INF treaty tends to bear this out. Also, weapons systems able to carry out these limited nuclear missions were usually "dual capable," and were therefore originally planned as general purpose forces. See Enthoven and Smith, *How Much is Enough?* pp. 125-132, and James A. Thompson, "Nuclear Weapons in Europe: Planning for NATO's Nuclear Deterrent in the 1980s and 1990s," *Survival*, Vol. XXV, No. 3 (May/June 1983), pp. 98-109.

8. See Ralph Sanders, *The Politics of Defense Analysis* (New York: Dunellen, 1973), especially Chapter 3, and Laurence E. Lynn and Richard I. Smith, "Can the Secretary of Defense Make a Difference?" *International Security* 7 (Summer 1982), pp. 45-69.

9. Lawrence J. Korb, "On Making the System Work," in Philip S. Kronenberg, ed., *Planning US Security* (Washington: National Defense University, 1981), pp. 139-146. Bob Kennedy has pointed out to me that short-term interests are also important. The problem is that in choosing options to resolve short-term issues, the system fails to provide a long-term guide to evaluate those options. Only with a long-term framework can the future value of present decisions be assessed.

2. Planning Strategic Nuclear Forces

1. See Desmond Ball, "Countervailing Targeting: How New? How Viable?" Strategic and Defense Studies Centre, Australian National University, 1980. See also Kevin Lewis, "US Strategic Force Planning: Restoring the Links between Strategy and Capabilities" Santa Monica: RAND P-6742 (January 1982) for an argument that force planning decisions have been separated from employment planning.

2. Bernard Brodie, ed., *The Absolute Weapon* (New York: Harcourt Brace and Company, 1946). See especially Brodie's chapter II, "Implications for Military Posture," pp. 70-107.

3. See David Alan Rosenberg, "The Origins of Overkill," *International Security* (Spring 1983), pp. 3-71, 140-162.

4. *Ibid*. See also Rosenberg, "A Smoking Radiating Ruin at the End of Two Hours: Documents on American Plans for Nuclear War with the Soviet Union, 1954-1955." *International Security* (Winter 1981-82), pp. 3-38.

5. *Ibid*.

6. Alain Enthoven and K. Wayne Smith, *How Much is Enough?* (New York: Harper, 1962), p. 167.

7. See Fred M. Kaplan's *Wizards of Armageddon* (New York: Simon and Schuster, 1983) for an interesting look at the personalities and players who fashioned American nuclear strategies and forces.

8. Enthoven and Smith, *How Much is Enough?* p. 170.

9. *Ibid*.

10. See Robert Sheer, "How the Nuclear Arms Race may have Started," *Denver Post*, April 11, 1982, p. B1.

11. Serious students of nuclear exchange calculations will not be satisfied with this rudimentary explanation. They are referred to John A. Battilega and Judith K. Grange, eds. *The Military Application of Modeling* (Wright-Patterson AFB: AFIT), Chapter 12. See also its companion volume, particularly Chapter 7, of *Case Studies in Military Application of Modeling* by Frances P. Hoeber published in 1981 by the Military Operations Research Society.

12. Thus to meet new targeting guidance, an aim point which had been directly over the Kremlin was moved to halfway between the Kremlin and an electric power station located one mile to the south. See Rosenberg, "The Origins of Overkill," p. 17.

13. McNamara's FY 1963 Posture Statement, a speech in Chicago in February 1962, and his Athens speech the same year outlined his thinking on counterforce strategy. The most famous articulation of counterforce policy is in his June 1962 speech at Ann Arbor. See William W. Kaufmann, *The McNamara Strategy* (New York: Harper and Row, 1964), pp. 74-75 and Desmond Ball "Targeting for Strategic Deterrence," London: International Institute for Strategic Studies (Adelphi Paper 185), Summer 1983.

14. Enthoven and Smith, *How Much is Enough?* p. 195.

15. Samuel Glasstone and Philip J. Dolan, eds., *The Effects of Nuclear Weapons* (US Department of Defense), Washington: GPO, 1977.

16. *Ibid.*, pp. 6-7.

17. Author's calculations. See Enthoven and Smith, pp. 180-182. CEP—Circular Error Probable—is conventionally defined as the radius of a circle, centered on a target, in which one-half of the bombs will fall.

18. See Stanley Sienkiewicz, "Observations on the Impact of Uncertainty in Strategic Analysis" in John F. Reichart and Steven R. Sturm, *American Defense Policy* (Baltimore: Johns Hopkins, 1982), pp. 217-227.

19. As Christopher Branch explains it, this approach asks, "If the same war were fought many times, what outcome could we expect to occur most frequently?" These methods of dynamic analysis are used by the Defense Department to support both guidance and policy. See Branch, *Fighting a Long Nuclear War* (Washington: National Defense University Press, 1984), pp. 59-65.

20. See Donald M. Snow, *The Nuclear Future* (The University of Alabama Press, 1983), pp. 150-156 for a good quantitative analysis of the targeter's dilemma. See also George V. Seiler, *Strategic Nuclear Force Requirements and Issues* (Maxwell AFB: Air University Press, 1983).

21. In keeping with the limited purposes and scope of this book, the presentation on kill probability, force exchange models, problems of "fratricide" and other important but complex calculations that enter into force planning decisions have been sidestepped. Those wishing to

delve more deeply into these issues should see Lynn Davis and Warner Schilling, "All You Ever Wanted to Know About MIRV and ICBM Calculations but Were not Cleared to Ask," *Journal of Conflict Resolution*, Vol. 17, No. 2 (June 1973) and John D. Steinbrunner and Thomas M. Garwin, "Strategic Vulnerability: The Balance Between Prudence and Paranoia," *International Security*, Vol. 1. No. 1 (Summer 1976), pp. 138-181.

22. Enthoven and Smith, *How Much is Enough?* p. 171. The Navy projected that a fleet of 45 submarines, with 29 deployed at all times, could destroy 232 Soviet targets, "which was sufficient to destroy all of Russia." See Rosenberg, "The Origins of Overkill," p. 57.

23. *Ibid.*, p. 178.

24. Desmond Ball, *Politics and Force Levels* (Berkeley: University of California Press, 1980), p. 62. See also Rosenberg, "The Origins of Overkill."

25. *Ibid.*, p. 68.

26. *Ibid.*, p. 75.

27. *Ibid.*, p. 212-231.

28. Colonel Glenn A. Kent, "On the Interaction of Opposing Forces under Possible Arms Agreements," (Cambridge: Harvard University Center for International Affairs. Occasional Papers in International Affairs, No. 5, March 1963). More famous was Kent's study that helped convince McNamara that "damage-limiting" was not an affordable strategy. See Kaplan, *Wizards of Armageddon*, pp. 321-325.

29. Lawrence Freedman, *The Evolution of Nuclear Strategy* (New York: St. Martin's Press, 1983), p. 340.

30. *Ibid.*

31. The following relies on Desmond Ball's explication of the Carter administration's decision on MX basing in Reichart and Sturm, eds., *American Defense Policy*, pp. 234-239.

32. This account relies on Strobe Talbott's *Deadly Gambits* (New York: Knopf, 1974), pp. 259-262.

33. Desmond Ball, "Counterforce Targeting: How New? How Viable?" in Reichart and Sturm, eds., *American Defense Policy*, pp. 234.

34. Talbott, *Deadly Gambits*, pp. 343-352. At this writing it appears that conclusion may be reversed by the end of the second term. A ratified INF Treaty is in hand, and the prospects for a Vladivostok-like START framework appear good.

35. Freedman, *The Evolution of Nuclear Strategy*, p. 400.

36. In a time of attempts to move away from reliance on assured destruction to a system of strategic defense, it is important to emphasize that quantitative assessments of the strategic balance cannot easily be cast aside. While one may wish to begin to construct a strategic defense that no longer compels the United States to rely solely on an adversary's rational decision to avoid unacceptable damage, the path towards such a capability is strewn with uncertainty and instability. Calculations of the strategic balance will continue to be important along the way. For one development of this argument, see Glenn A. Kent and Randall J. DeValk, *Strategic Defenses and the Transition to Assured Survival* (Santa Monica: RAND, 1986).

3. Planning General Purpose Forces

1. See Terry L. Deibel, *Commitment in American Foreign Policy* (Washington: National Defense University, 1980).

2. Enthoven and Smith, *How Much is Enough?* p. 211.

3. Harold Brown, *Department of Defense Annual Report FY 1980*, p. 63.

4. These factors are drawn from a number of sources, but a similar list is developed by Leslie Gelb and Arnold Kuzmack. See "General Purpose Forces" in Henry Owen, ed., *The Next Phase in Foreign Policy* (Washington: Brookings, 1973), pp. 213-217.

5. Including active and reserve forces. See the in-depth examination of this force planning by Paul Hammond in Warner Schilling, Paul Hammond, and Glenn Snyder, *Strategy, Politics and Defense Budgets* (New York: Columbia, 1962), pp. 298-378.

6. William W. Kaufmann, *Planning Conventional Forces 1950-1980* (Washington: Brookings, 1982), p. 3.

7. *Ibid.*, pp. 6-7.

8. Defense Secretary Laird noted in his 1970 Annual Report that force levels to support the "2½ war principle" were never achieved. Melvin R. Laird, *DOD Annual Report, FY 1971*, pp. 10-11.

9. "Schlesinger Seeks more Army Divisions," *Washington Post*, October 17, 1975, p. 6.

10. As in the case of the modeling of strategic weapons exchanges, the quantitative detail in this primer on force planning will be more suggestive than complete. Those desiring to study general purpose force modeling in greater depth should consult Battilega and Grange, *The Military Application of Modeling*, Chapters 6, 7, and 8, and Hoeber, *Application of Modeling*, Chapters 1, 5, and 6. See also William Kaufmann, "The Arithmetic of Force Planning," in Steinbruner and Sigal, eds., *Alliance Security: NATO and the No-First-Use Question* (Washington: Brookings, 1983), and William P. Mako, *US Ground Forces and the Defense of Central Europe* (Washington: Brookings, 1983), especially Appendix A. An important critique of Kaufmann's

use of Lanchester equations for force planning can be found in the Summer 1987 issue of *International Security*.

11. Again this planning guidance is drawn from a number of sources, but it relies most heavily on the points presented in the Congressional Budget Office (CBO) paper "Planning US General Purpose Forces: Army Procurement Issues" (Washington: CBO, 1976), pp. 5-9. See also "Planning US General Purpose Forces: Overview" (Washington: CBO, January 1976), pp. 8-11, and Kaufmann, *Planning Conventional Forces.*.

12. One of the most notable critiques, and one which also contains valuable explanations of quantitative methods used in force planning, is the *Comptroller General's Report to the Congress*: "Models, Data and War: A Critique of the Foundation for Defense Analyses" (Washington: GAO, March 12, 1980). See also R. James Woolsey. "Planning a Navy: The Risks of Conventional Wisdom," *International Security* (Summer 1978), pp. 17-29.

13. For a more complete development of these and other factors see Robert Lucas Fischer, *Defending the Central Front: The Balance of Forces* (London: IISS, autumn 1976,) Adelphi Paper 127, pp. 16-35.

14. See, for example, J. A. Stockfisch, *Models, Data and War: A Critique of the Study of Conventional Forces* (Santa Monica: RAND R-1526-PR, March 1975), pp. 24-25.

15. For a good depiction of how optimistic or pessimistic assumptions can skew the force balance see James Blaker and Andrew Hamilton, *Assessing the NATO/Warsaw Pact Military Balance* (Washington: CBO, December 1977), Stockfisch, *Models, Data and War*, p. 25, and Mako, *US Ground Forces and the Defense of Central Europe*, Appendix B and C.

16. See the GAO's report "A Critique of the Foundation for Defense Analyses," pp. 62-66.

17. Blaker and Hamilton, *Assessing the Pact Balance*, p. 14. For an opposing view of quantitative relationships among firepower potential, force ratio, and troop movement see Colonel T. N. Dupuy's book, *The Evaluation of Weapons and Warfare*, and other works.

18. *Ibid.*

19. Stockfisch, *Models, Data and War*, pp. 12-14.

20. See "A Critique of the Foundation for Defense Analyses," pp. 66-70.

21. The developmental chapter from Lanchester's *Aircraft in Warfare* is reprinted in James R. Newman, ed., *The World of Mathematics*

(New York: Simon and Schuster, 1965), Vol. 4, pp. 2138-2159. See also "A Critique of the Foundation for Defense Analyses," pp. 66-68. James G. Taylor has catalogued the enrichment of the Lanchester models in great detail and, in addition, has provided us with a comprehensive bibliography of the Lanchester theory of combat. See his two-volume work, *Lanchester Models of Warfare* (Monterey: Naval Postgraduate School, 1983). Also see Joshua Epstein's dynamic analysis without Lanchester theory, *The Calculus of Conventional War* (Washington: Brookings, 1985). Finally, see John W. R. Leppingwell, "The Laws of Combat? Lanchester Reexamined" *(International Security*, Summer 1987), Vol. 12, No. 1, pp. 89-134.

22. The limits of the square law define the issue of "force to space ratio." Across a given area of forces in contact, one can meaningfully concentrate only a limited force without yielding to the defender advantages in command and control. Mass beyond a certain point is subject to diminishing returns. In *The Calculus of Conventional War*, Epstein notes that impressive evidence exists to contradict the Lanchester square equations. See also Leppingwell, "The Laws of Combat."

23. Enthoven and Smith, *How Much is Enough?* p. 119.

24. *Ibid.*, pp. 132-142.

25. See Blaker and Hamilton, *Assessing the Pact Balance*, for the ways in which assumptions can alter this balance and call for greater (or smaller) US troop deployments.

26. Derived from Blaker and Hamilton, *Assessing the Pact Balance*, Fischer, *Defending the Central Front*, and Kaufmann, "Nonnuclear Deterrence" in Steinbrunner and Sigal, *Alliance Security*.

27. These assumptions held by CBO force planners may not be shared by all. See "US Tactical Air Forces: Overview and Alternative Forces, FY 1976-1981" (Washington: CBO, April 14, 1976), p. 7.

28. See Air Force Manual 1-1, "Basic Aerospace Doctrine of the United States Air Force," 16 March 1984.

29. "Planning US General Purpose Forces: Overview" (Washington: CBO, January 1976), p. 23.

30. Although the missions of the tactical air forces are generally agreed upon, the priorities allocated to them are not. See William D. White, *US Tactical Air Power* (Washington: Brookings, 1974), pp. 61-73. White also includes the tactical missions of reconnaissance, electronic warfare, and tactical airlift. For our purposes, the size of the

recce/ECM force will be determined by the total number of tactical fighter wings. Airlift requirements will be discussed in Chapter 4.

31. Although the simulation was far from perfect, in the Lebanon conflict in 1982 Syrian pilots flying front line Soviet aircraft lost a total of 86 aircraft to an unscratched Israeli Air Force flying primarily US aircraft. Qualitative factors are clearly at work here. Those calling for greater numbers of less sophisticated aircraft appear from this to have a weakened argument.

32. Derived from the reconstruction of the 1962 Force Planning Study. See Kaufmann, *Planning Conventional Forces*, p. 9.

33. "Planning US General Purpose Forces: The Tactical Air Forces" (Washington: CBO, July 1977), pp. 14-15.

34. See "Deep Attack in Defense of Central Europe: Implications for Strategy and Doctrine," in *Essays on Strategy* (Washington: NDU Press, 1984), pp. 29-75.

35. William D. White, *US Tactical Air Power*, generated this mission distribution of Air Force tactical air support by examining the operational theaters of Europe and the Pacific in World War II as well as in the Korean War. While the proportions held true during the land-intensive campaigns in Europe and Korea, counter air sorties dominated the Pacific theater. In Vietnam interdiction sorties consumed close to 90 percent of the tactical air sorties. See White, *US Tactical Air Power*, p. 68. In Europe, the ideal is to have the majority of aircraft capable of "swinging" from CAS to BAI to air-to-air.

36. Based on the recent deployment to Europe, and restricting the A-10 to CAS and the F-15 to air superiority, the percentages in terms of squadrons deployed equates to 20 percent CAS, 65 percent interdiction, and 15 percent air superiority. But the majority of F-16/F4 in the interdiction role can accomplish both CAS and counter air missions. Clearly these percentages will change with the conduct of the campaign, the tactics of the air component commander, and increasingly sophisticated aircraft/systems such as the F-15E, F-16C/D, LANTIRN and AMRAAM.

37. The following is based on a discussion comparing firepower alternatives found in "US Air and Ground Conventional Forces for NATO: Firepower Issues" (Washington: CBO March 1978), pp. 17-29.

38. *Ibid.*, p. 33. LANTIRN on the F-16 and F-15E will provide the US a considerable night/adverse weather advantage for battlefield interdiction. A follow-on CAS aircraft with this capability is not planned at this writing, but an "A-16" would bring flexibility and survivability to the close air support/battlefield air interdiction mission.

39. Joshua M. Epstein, *Measuring Military Power: The Soviet Air Threat to Europe* (Princeton: Princeton University Press, 1984).

40. *Ibid.*, p. 172. Epstein notes that a "worst plausible case" based on analysis is considerably different from and is worth much more to the force planner than a "worst possible case."

41. *Ibid.*, p. 81.

42. The following discussion of naval strategy, scenarios, and missions relies primarily on "Planning US General Purpose Forces: The Navy" (Washington: CBO, December 1976) pp. 3-15, and "Shaping the General Purpose Navy of the Eighties: Issues for FY 1981-1985" (Washington: CBO, January 1980), pp. XIII-XX. See also William Kaufman, *A Thoroughly Efficient Navy* (Washington: Brookings, 1987). Strategic sealift will be discussed in the following chapter.

43. See "US Naval Forces: The Peacetime Presence Mission" (Washington: CBO, December 1978).

44. Of course, carrier-based aircraft perform in a defensive role for the task force as well. Adding defensive systems can also complicate carrier operations. The Navy has suggested two Aegis carrier escorts for each carrier to obviate these ships being forced to fire across the carrier to defend it.

45. See the development of the carrier task force concept in "Planning US General Purpose Forces: The Navy," (Washington: CBO, December 1976), pp. 17-30.

46. Totals based on two Aegis per carrier and a total of nine auxiliary ships—support ships added to the total fleet based on a factor of 10 percent, including maintenance and repair ships and tenders.

47. "United States Maritime Strategy and the Naval Shipbuilding Program" (Washington: DOD, August 1976). Also see "Planning US General Purpose Forces: The Navy," (Washington: CBO, December 1976), pp. 52-55. Since 1975 the Navy has considered 600 ships the minimum number required to meet its varied missions. Defense Secretary Brown announced in 1978 that a 525-ship fleet would be adequate to meet these missions in the 1980s, based on 12 aircraft carriers. See Samuel F. Wells, Jr., "A Question of Priorities: A Comparison of the Carter and Reagan Defense Programs," *Orbis*, Vol. 27, No. 3 (Fall 1983), pp. 641-666.

48. Former Navy Secretary Lehman has justified the expansion of the Navy by rejecting a passive defensive strategy centered on the GIUK gap and has spoken often of the necessity of taking the battle to the

Soviets. See Major Hugh K. O'Donnell, Jr., USMC, "Northern Flank Maritime Offensive," *Proceedings* (September 1985), p. 42. See also Robert W. Komer, "Maritime Strategy vs. Coalition Defense," *Foreign Affairs* (Summer 1982), pp. 1124-1144, William Kaufmann, *A Reasonable Defense* (Washington: Brookings, 1985), and the Mearsheimer-Brooks debate over naval strategy in the Fall 1986 issue of *International Security*.

49. William Kaufmann, *The 1986 Defense Budget* (Washington: Brookings Institution, 1985), p. 35.

50. *Ibid.*

51. See Office of the Joint Chiefs of Staff, *Military Posture FY 1986*, pp. 73-74.

52. Kaufmann, *The 1986 Defense Budget* (Washington: Brookings 1985), p. 35. Kaufmann generates a somewhat different force in more recent work. See *A Thoroughly Efficient Navy* (Washington: Brookings, 1987).

4. Planning Rapidly Deployable Forces

1. Quoted in Samuel Huntington, *The Common Defense* (New York: Columbia University Press, 1961), p. 27.

2. These goals are extracted from the *Department of Defense Annual Report* for FY 1979 and 1980. See also Albert Wohlstetter, "Meeting the Threat in the Persian Gulf," *Survey* (Spring 1980), pp. 128-188.

3. This chapter is the distillation of another work by the author. See Robert P. Haffa, Jr., *The Half War: Planning US Rapid Deployment Forces to Meet a Limited Contingency, 1960-1983* (Boulder: Westview, 1984).

4. *DOD Annual Report FY 1980*, p. 99.

5. Testimony of Under Secretary of Defense Robert Komer, US Congress, House, Committee on Armed Services, DOD Authorization for Appropriations for FY 1981, *Hearings on Military Posture and H. R. 6495*, 96th Congress, 2nd Session (Washington, DC: United States Government Printing Office, 1980), part 1, p. 457. Despite these realizations, the Carter administration "took the position that the capabilities already incorporated in the force structure for a minor contingency (approximately three divisions and five tactical fighter wings) were sufficient to cope with an emergency in the Gulf." See William W. Kaufmann, "The Defense Budget" in Joseph A. Pechman, ed., *Setting National Priorities: The 1982 Budget* (Washington, DC: Brookings, 1981), p. 153ff.

6. *Strategic Survey 1979* (London: International Institute for Strategic Studies, 1980), p. 1.

7. If the purpose of the force was to lend American assistance to stabilize the political situation in the Western-oriented Gulf states, the composition of the RDJTF might be limited to the traditional light deployment units, particularly the Marine Corps. If the purpose of the force was to defend against Soviet intervention, a multi-service force of significant capability was required. In either case, too visible a link with the United States might weaken the regimes in question in a domino-like extension of the Iranian revolution. Thus military "presence" remained an inadequate solution in the immediate area, and the

requirement for an "over the horizon" force demanded a rapid deployment capability. This debate was addressed in the *Hearings on the RDF before the Senate Armed Services Subcommittee on Seapower and Force Projection*, March 12, 1981.

8. These conflicts included a quarrel between the Air Force and the Marines over the control of airspace in the objective area, questions regarding responsibilities of the Navy-directed Pacific Command in the Indian Ocean area, and doubts expressed about the future of the Marine Corps amphibious and rapid deployment missions, if not the future of the Corps itself.

9. Enthoven and Smith, *How Much is Enough?* p. 237.

10. Marshall Foch prior to 1914 stated that he needed only one British soldier in France, provided it could be guaranteed that the man would be killed in action on the first day of the war. Few today would advocate such a singular deterrent to aggression, but tripwire theories spring from this concept and offer equally low levels of confidence.

11. Enthoven and Smith, *How Much is Enough?* p. 234.

12. *Ibid.*

13. *Ibid.*, p. 235.

14. See Richard B. Rainey, Jr., "Mobility—Airlift, Sealift and Pre-positioning" (Santa Monica: RAND 1966), P3303.

15. Enthoven and Smith, *How Much is Enough?* p. 236.

16. Frances P. Hoeber, *Case Studies in Military Applications of Modeling* (Arlington, VA: Hoeber Corp., May 1980), p. III-4.

17. Bernard Brodie, *A Guide to Naval Strategy* (Princeton: Princeton University Press, 1958), p. 252.

18. Alain Enthoven, "The Role of Systems Analysis in the FDL," included in testimony in US Congress, House, Committee on Armed Services, *Hearings on Military Posture for FY 1969*, 90th Congress, 2d session (Washington, DC: GPO, 1968), p. 8913.

19. US Congress, Senate, Committee on Armed Services, *Hearings on Department of Defense Appropriations for FY 1969 and H.R. 18707*, 90th Congress, 2d session (Washington, DC: GPO, 1968), p. 538.

20. US Congress, House, Committee on Armed Services, *Report of the Special Subcommittee on National Airlift* (Washington, DC: GPO, 1960), p. 4027.

21. *Ibid.*, p. 4051.

22. "SecDef Reports on National Military Airlift Plans." See also "Limited War Forces Stressed by Kennedy," *Washington Evening Star*, January 30, 1961.

23. *Air Force Review of the C-5A Program*, Department of the Air Force, July 1969, p. I-2.

24. US Air Force, Historical Division Liaison Office, *Strengthening USAF Airlift Forces, 1961-1964* (Washington, DC: Historical Division, February 1966), pp. 4-5.

25. Enthoven and Smith, *How Much is Enough?* p. 237.

26. See Walter R. Shope, "The Lessons of Nifty Nugget," *Defense 80* (December 1980), pp. 14-22, and John J. Fialka, "The Grim Lessons of Nifty Nugget," *Army* (April 1980), p. 40. Shope notes that one of the most significant discoveries in the course of the exercise was the separation between operational contingency planning and resource allocation. Fialka concentrates on the results, specifically the reforms and shortfalls. See also US Department of Defense, "An Evaluation Report on Mobilization and Deployment Capability Based on Exercise Nifty Nugget-78" (Washington, DC: DOD, 30 June 1980). Most recently, the lessons of Nifty Nugget for the new Unified Transportation Command were spelled out by Peter Grier in *Military Logistics Forum*, Vol. 3, No. 9 (June 1987), pp. 42-43.

27. Compiled from the *DOD Annual Reports* for FY 1980, 1981, and 1982.

5. Planning US Forces

1. See Warner R. Schilling, "US Strategic Nuclear Concepts in the 1970s," *International Security* (Fall 1981), pp. 48-79.

2. *Ibid.*

3. The criterion for assured destruction also remained the same. The percentages of population and industrial damage equating to assured destruction were essentially the same in 1979 as they were in 1965. See Defense Secretary Brown's FY 1979 Posture Statement, pp. 49-55.

4. Another interpretation is that MX acquisition has been driven by the need for increased hard-target kill capability, and that it is "warfighting" that has triumphed over "assured destruction" concepts of deterrence rather than "perceived equality."

5. For expansion on some of these points see Kevin N. Lewis, "US Strategic Force Planning: Restoring the links between Strategy and Capabilities," Santa Monica: RAND P-6742 (January 1982).

6. *Ibid.* Lewis notes that the original B-1 buy of 240 equated to the B-52 G and H models in the strategic offensive role. Two hundred MX with 10 RVs approximated the Minuteman force. Twenty-five Trident submarines, with higher on-station rates, came close to the number of alert tubes in the Polaris/Poseidon fleet.

7. George Kistiakowsky's report is quoted from Desmond Ball's, "Targeting for Strategic Deterrence," Adelphi Paper Number 185, London: International Institute for Strategic Studies, p. 41.

8. *Ibid.* p. 37.

9. Thomas Powers, "Choosing a Strategy for World War III," *The Atlantic Monthly* (November 1982), p. 95. But see Scott D. Sagan, "SIOP-62: The Nuclear War Plan Briefing to President Kennedy," *International Security* (Summer 1987), Vol. 12, No. 1, pp. 22-51.

10. Ball, "Targeting for Strategic Deterrence," p. 41.

11. For one image of what such a strategic force might look like, see Joshua Epstein, *The 1988 Defense Budget* (Washington: Brookings, 1987), pp. 15-32.

12. See Caspar W. Weinberger, *Annual Report FY 1983*, p. I-15.

13. For early formulations of the horizontal escalation concept see Fred Iklé, "The Reagan Defense Program: A Focus on the Strategic Imperative," *Strategic Review* (Spring 1982), pp. 11-34, and also his "Strategic Principles of the Reagan Administration," *Strategic Review* (Fall 1983), pp. 13-18.

14. Certainly such a thesis is supported by the staff study emanating from the Senate Armed Services Committee led by defense stalwarts Senator Nunn and former Senator Goldwater. See *Armed Forces Journal*, October 1985 (extra issue), and the study itself, "Defense Organization: The Need for Change." The study formed the basis for the subsequent Goldwater-Nichols legislation.

15. The 1985 study, "Toward a More Effective Defense," was conducted by Georgetown University's Center for Strategic and International Studies. See Michael Weisskopf, "Defense Buying Systems and Command Faulted," *Washington Post*, February 26, 1985, p. A7. For opposing views on JCS reform see John G. Kester, "The Role of the Joint Chiefs of Staff" in Reichart and Sturm, eds., *American Defense Policy*, pp. 527-545, and Victor H. Krulak, *Organization for National Security*, Washington: US Strategic Institute, 1983.

16. By 1985, the Army was planning for five light divisions. Early plans for significant manpower increases were dropped in 1984 and have not been returned to. See "Army Reported Ready to seek a New Division," *Philadelphia Inquirer*, December 24, 1983, p. 8, and Walter Andrews, "Defense approves Army's plan to form 5 light infantry divisions," *Washington Times*, August 23, 1985, p. 5. An attendant problem is the dependence on civilian crews, particularly in the Navy, as force structure expands without comparable manpower growth. See Fred Hiatt, "Civilian Officers' Strike Immobilized Navy Ship," *Washington Post*, April 7, 1985, p. A5.

17. General John A. Wickham, Jr., "White Paper on the Light Division," 16 April 1984.

18. See "Deep Attack in Defense of Central Europe: Implications for Strategy and Doctrine" in *Essays on Strategy* (Washington: National Defense University Press, 1984), pp. 29-75.

19. "Army and Air Force Sign Agreement" OSD News Release, May 22, 1984.

20. See William W. Kaufmann, *The 1986 Defense Budget* (Washington: Brookings, 1985), p. 35.

21. Robert Komer, "Maritime Strategy vs. Coalition Defense," *Foreign Affairs*, Summer 1982, pp. 1124-1144. For an argument that

there lies a middle ground between the two strategies see Keith A. Dunn and William D. Staudenmaier, ''Strategy for Survival,'' *Foreign Policy* No. 52 (Fall 1983), pp. 22-41.

22. Robert Komer, ''The Neglect of Strategy,'' *Air Force* (March 1984), pp. 51-59.

23. A glimpse of the political magnitude of the task of DOD reorganization was revealed by its architect in Archie D. Barrett, *Reappraising Defense Organization* (Washington: National Defense University, 1983), pp. 279-284.

24. Richard L. Kugler, ''Whither Defense Analysis: Toward a New Gestalt,'' May 1984 (unpublished).

25. See Thomas L. McNaugher, ''Balancing Soviet Power in the Persian Gulf,'' *The Brookings Review* (Summer 1983), pp. 20-24.

26. Caspar W. Weinberger, *Annual Report FY 1983*, p. II-103.

27. Caspar W. Weinberger, *Annual Report FY 1984*, p. 194.

28. See Bernard E. Trainor, ''Concern Reported in U.S. Military on Gulf Command Structure,'' *New York Times*, August 15, 1987, p. 3.

29. James A. Russell, ''Deployment: Will TRANSCOM Make a Difference?'' *Military Logistics Forum* (Vol. 3, No. 9, June 1987), p. 39.

30. See P. M. Dadant, *Improving U.S. Capability to Deploy Ground Forces to Southwest Asia in the 1990s* (Santa Monica: RAND, February 1983).

31. Scott C. Truver, ''Sealift Manning: Critical Period, Critical Choices,'' *Armed Forces Journal International* (July 1987), p. 34.

32. Interview with Vice Admiral Walter T. Piotti, Commander, Military Sealift Command, *Armed Forces Journal International* (July 1987), pp. 48-52.

33. See Jeffrey Record, *U.S. Strategic Airlift: Requirements and Capabilities*, National Security Paper No. 2, Washington: Institute for Foreign Policy Analysis, 1985. Preliminary results of the JCS Revised Intertheater Mobility Study (RIMS) suggest that the airlift requirements reflected in the 1981 Congressionally Mandated Mobility Study will not diminish.

34. Caspar W. Weinberger, *Annual Report to Congress FY 1988*, p. 230.

35. See Patrick E. Tyler, ''Kuwait May Offer Support Facilities,'' *Washington Post*, July 21, 1987 p. A1, and John H. Cushman, Jr.,

"Weinberger Urges Bases in Gulf Area for U.S. Air Patrol," *New York Times*, May 25, 1987, p. 1.

36. MacLaury, in his introduction to Joshua Epstein's *The 1988 Defense Budget*, Washington, Brookings, 1987, p. vii.

INDEX

151

ABOUT THE AUTHOR

Colonel Robert P. Haffa, Jr. United States Air Force, directs a Staff Group for the Chief of Staff of the Air Force, in Washington, DC. He wrote this book while a Senior Fellow at the National Defense University.

Colonel Haffa was educated at the United States Air Force Academy (BS,) Georgetown University (MA), the University of Arkansas (MS), and the Massachusetts Institute of Technology (Ph.D). At the Air Force Academy, he designed and taught courses on American Defense Policy and Strategy and Arms Control, and later was Professor and Acting Head of the Department of Political Science. Other Air Force assignments have included operational tours in Vietnam, the United Kingdom, and the Republic of Korea as a weapons system operator in the F-4 aircraft; operational plans officer and the Chief of Training in an Air Force tactical fighter wing.

His articles, "An Introduction to Arms Control" and "An Analytical Approach to Bureaucratic Politics," were included in Endicott and Stafford, eds., *American Foreign Policy* (1977). He also co-authored the article "Supply-Side Nonproliferation" in *Foreign Policy* 42 (Spring 1981). He is the author of the book, *The Half War: Planning US Rapid Deployment Forces to Meet a Limited Contingency.*

Colonel Haffa's professional military education includes the Squadron Officer School, Air Command and Staff College, and the National War College.

Rational Methods, Prudent Choices:
PLANNING US FORCES

Text and display lines composed in Times Roman
Book design by Donald Schmoldt
Cover artwork and figures prepared by Rhonda Story and
Juan Medrano

NDU Press Editors: Thomas Gill and Donald Schmoldt
Editorial Clerk: Carol Valentine